Seed Production of Field Crops

NIPA® GENX ELECTRONIC RESOURCES & SOLUTIONS P. LTD.
New Delhi-110 034

About Authors

Dr. Shyam Sundar Mondal, is the Professor and former Head of the Department, Agronomy, Bidhan Chandra Krishi Viswavidyalaya, Mohanpur, Nadia, W.B. Dr. Mondal has got more than 33 years of experience in agricultural teaching, research and extension. He has received several honours and awards including Honorary Fellow of Indian Society of Agronomy and West Bengal Academy of Science and Technology, India. He also obtained 'Best Paper Award' conferred by Indian Potash Institute. He has guided 25 M.Sc. and 20 Ph.D. students. He has more than 250 research publications in referred national and international journals.

Dr. Mithun Saha, a young promising agriculture technologist, is currently serving Department of Agriculture, Government of West Bengal as an Agricultural Development Officer. Dr. Saha was engaged in seed production research on different crops at Bidhan Chandra Krishi Viswavidyalaya, W.B. during his Ph.D. studies. He posses an excellent academic record. He received National Scholarship and award of Jawaharlal Nehru Scholarship for Ph. D. study. Dr. Saha has already published 25 research papers in referred national and international journals.

Dr. Kajal Sengupta, Professor, Department of Agronomy, Bidhan Chandra Krishi Viswavidyalay, Mohanpur, Nadia, West Bengal, is an eminent agronomist, engage in research, teaching and extension activity. Dr. Sengupta shares his twenty five years long experience in this book. He was member of several referred committee viz. Council of Vocational Education and Training. He guided fifteen M.Sc.and five Ph.D. students in Agronomy. He was Project In charge sponsored by many Government and Private Sector.

Dr. Hirak Banerjee obtained B.Sc (Ag.) Hons., M.Sc (Ag.) and Ph.D degrees in Agronomy from Bidhan Chandra Krishi Viswavidyalaya, Mohanpur, West Bengal. He was awarded with 'Certificate of Merit' and 'Gold Medal' in M.Sc. (Ag) for his outstanding performance. He started his professional career as Subject Matter Specialist (Agronomy) in Nadia Krishi Vigyan Kendra in 2005. Later, he joined as Assistant Professor in the Department of Agronomy, Bidhan Chandra Krishi Viswavidyalaya in 2006. He has completed more than 17 years of work experience in teaching, research and extension programmes. He received 'Young Scientist' award in 2013 and 'Fellow' award in 2018 from 'Crop and Weed Science Society', India. Research areas of his interest are agronomy, crop nutrition, weed science and agro-forestry. He has guided 3 M.Sc. (Ag.) and 3 Ph.D students for the award of their degrees. He has published about 127 research papers in referred journals of National and International repute.

2nd Revised and Enlarged Edition

Seed Production of Field Crops

Shyam Sundar Mondal
Former Head and Professor
Department of Agronomy
Bidhan Chandra Krishi Viswavidyalaya
Mohanpur, Nadia, West Bengal

Mithun Saha
Assistant Director of Agriculture
Department of Agriculture
Govt. of West Bengal, West Bengal

Kajal Sengupta
Former Dean, Faculty of Agriculture
and Head & Professor
Department of Agronomy
Bidhan Chandra Krishi Viswavidyalaya
Mohanpur, Nadia, West Bengal

Hirak Banerjee
Associate Professor
Department of Agronomy
Bidhan Chandra Krishi Viswavidyalaya
Mohanpur, Nadia, West Bengal
Ex-Subject Matter Specialist (Agronomy)
Nadia KVK, Nadia, West Bengal

nipa®

NIPA® GENX ELECTRONIC RESOURCES & SOLUTIONS P. LTD.
New Delhi-110 034

NIPA® GENX ELECTRONIC RESOURCES & SOLUTIONS P. LTD.

101,103, Vikas Surya Plaza, CU Block
L.S.C.Market, Pitam Pura, New Delhi-110 034
Ph : +91 11 27341616, 27341717, 27341718
E-mail:newindiapublishingagency@gmail.com
www: www.nipabooks.com

For customer assistance, please contact
Phone: + 91-11-27 34 17 17
Fax: + 91-11-27 34 16 16
E-Mail: feedbacks@nipabooks.com

ISBN: 978-93-94490-78-9

Composed and Designed by NIPA®.

Preface to the First Edition

India is the homeland of diversified crops. The varying agro-ecological condition favours the natural growth and production of these crops. But it needs to be harvested to their maximum potentiality. Our work is to restore the type or variety suitable for particular region and scientific production of the seed or planting materials for further use. Sustained growth to cope with increasing demand would depend more on the pace of development and adoption of innovative technologies, the seed would continue to be vital component for decades to come. Though the history of seed production is old, the concept has received only limited attention from the scientific point of view and that only in relatively recent time with the development of seed industry. As a consequence, the key player for transforming traditional farming into market led and demand driven and value added farming is seed industry. The private seed industry is no more confined to just production and marketing of seed. It has as well as acquired technological strength to cater to the varietals needs of tomorrow.

Several periodical and processing of the national and international workshops, seminars, conferences and few books have been published to disseminate knowledge in the fast growing discipline of seed science. But there is no such text as well as research publication which will help the researchers and students to congregate the information in integrated manner.

The authors acknowledge their indebtedness to the authors of various books, monographs, periodicals from which they have drawn matter. They also greatly acknowledge to the books like Seed Technology by R.L.Agarwal; Hand Book of Agriculture by ICAR; Field Crops by Rajendra Prasad and scientific publication of Indian Potato Research Institute, Shimla, Central Research Institute for Jute Allied Fibre, Barrackpore, Central Rice Research Institute, Cuttack and Indian Council of Agricultural Research.

Authors do not claim perfection but at the same time they have made all possible efforts with their vast experiences in teaching, research and extension to make this book up to date and useful to students and technologists from all point of view.

Shyam Sundar Mondal
Mithun Saha
Kajal Sengupta

Preface to the Second Edition

During the last eleven years since the first edition of this book was published, good amount of valuable technological information of seed production of field crops have accumulated. When we took up the task of revising the first edition of the book to make it update it was really herculean task. Instead of recasting the entire book, some of the most important information dealing with new concept of the new crops has been added at the later part of the book.

We wish to thank the assistance given by our colleagues in the university who gave us quite a lot of information. Our grateful thanks are due to Bidhan Chandra Krishi Viswavidyalaya for providing opportunity and facilities for bringing out the valuable revised edition.

Shyam Sundar Mondal
Mithun Saha
Kajal Sengupta
Hirak Banerjee

Contents

1

General Seed Production Technology

Current Status of Seed Production

Organized seed production in India became possible with the formation of National Seed Corporation (NSC) during 1963, which paved the way for a sound seed industry to come as a consequence of the introduction of hybrid technology during 1961-65. The release of high yielding dwarf varieties of wheat and rice by the 1960's gave further impetus to the growth of the seed industry in the country. For the total area under cultivation of 26.7 m ha and 44.8 ha of wheat and rice respectively, the seed requirement for these two crops alone come to 32 lakh tonnes and 13 lakh tonnes respectively of which the varieties contribute about 98 per cent and the hybrids 2 per cent. Only about 10-15 per cent of the total requirement is met by the seed producers leaving the balance for the farmers to manage and the seed producers have to grab the opportunity to supply quality seed materials in order to increase the production. Many of the problems that impede production are low input supply, moisture stress (>65 per cent of the area) insect, disease and weeds (10-40 per cent loss) and also poor shelf-life. Hybrid technology have defined solution in several crop plants specially vegetables, cotton, rice etc. with the development of appropriate hybridization and seed production. In 1990-91, the area planted to 'bought seed' was about 10.35 per cent. The total 'bought seed' was estimated at around 5.91 lakh tonnes valued at Rs 679.80 crores. The area planted to 'bought seed' increased by 3 per cent and the market size expanded significantly and was estimated at 8.64 lakh tonnes value at Rs 2,249 crores. The trends too suggests that price of seed though important for adoption of new technology, is not a constraint, if the seed ensures proportionate gains through higher productivity and higher value. There are two national institutions *viz.* National Seed Corporation and State Farms Corporation of India besides 13 State Seed Corporations situated throughout India and 147 large scale private seed companies to cater the requirement of about 27 lakhs tonne of seeds worth about Rs 4,500 crores leaving to a gap to be bridged.

The constraints adhere to the seed production with the Indian agriculture are lack of proper support from the government, improper coordination between various seed producing associations in meeting the demand, lack of training facilities to producers, improper knowledge of advantage on using 'bought seed' in farmers field but above all it requires highly technical know-how, trained personnel and resource supply.

Quality seed alone contribute for nearly 15 per cent of the total production and if adequate quantities seeds are not made available in time there is every possibility of low production. There is an urgent need to produce the seeds in order to meet the seed demand to match the growing demand, in feeding the swelling population in the new millennium.

1. General Seed Production Technology

1.1 Seeds and Sowing

Seed is a matured ripened ovule with entact embryo. Agronomically seed is defined as any planting material used for sowing or planting of crops, which may be maize or wheat seed, sugarcane setts, potato tuber, banana suckers etc. Seed is the most vital and important input for agricultural production.

Quality of Seed

i) Germination percentage: Seeds should have high germination percentage, more than 80% germination.

ii) Genetic purity of seed: Seeds should have at least 90% genetic purity. (Genetic purity is the percentage of true seeds in the seed material).

iii) Seeds should be free from other inert materials: mixture of sand, soil, wood, stem, leaf, and stone.

iv) Seeds should be free from mixture of other seeds and varieties.

v) Seeds should be free from weed seeds.

vi) Seeds should be free from pest and disease organism.

Types of Seed

Four classes of pure seeds are recognized by the International Crop Improvement Association. They are –

i) Breeder or Nucleus Seed: Seeds produced and supplied by the breeder. This is the initial amount of pure seed of an improved variety and its genetic purity is 100% and is very limited quantity. It is produced under direct supervision of the main breeding station.

ii) Foundation seed: Seeds multiplied from breeder's seed in the supervision of Government Farm or Farms recommended by Government or organization like National Seed Corporation and that is so handled as to maintain specific genetic identity and purity. Foundation seed is obtained from seed multiplication farm. Any seed produced in Govt. multiplication farm is usually called as foundation seed. It is not so pure genetically or physically as like as the breeder seed and it's production is carried out under supervision of Research Assistant. Purity of foundation seed is 99.5%.

iii) Registered seed: Registered seed is produced from foundation seed or from registered seed. Registered seed is genetically pure and is used to produce certified seed or registered seed. It is usually produced by progressive farmers according to technical advice and supervision by NSC. Often registered seed is omitted and certified seed is produced directly from foundation seed. This is the general practice in India. It's purity is 99%.

iv) Certified seed: Certified seed is produced from foundation, registered or certified seed. This is known because it is certified by a seed certification agency, in this case State Seed Certification Agency. This is produced by progressive farmers according to standard seed production practices. To be certified, the seed must meet the prescribed requirements regarding purity and quality. Certified seed is used for commercial crop production by the farmers. In certified seeds genetic purity is 99%. It's production is generally by State Seed Corporation, but NSC also undertakes supervision of certified seed production, if required.

Seed Treatment

Before sowing seeds are subjected to special treatment for different purposes. They are for convenience in sowing, control of pests and diseases, enhancing germination and for inoculation of *Rhizobium* culture.

i) Convenience in Sowing

There are two types of fibres *viz.* fuzz and lint which adhere with cotton seeds. On account of these fibre seeds of cotton cling together and can not be sown through drills or broadcast. To remove this seeds are rubbed together with a paste made of wet dung and earth and dried. After this treatment, seeds can not adhere to each other but roll apart just like grain. The fuzz can also be removed by treating the seeds with concentrated sulphuric acid for two minutes, washed off free sulphuric acid, dried thoroughly and sown.

Before sowing the fruit of coriander split into two halves by rubbing lightly. In case of garlic the compound bulb has to be split into individual ones.

Due to very small size it is difficult to sow the small seeds like ragi, tobacco etc. So these seeds are mixed with earth and sand.

ii) Insect-Pest and Diseases Control

There are several pests and diseases which are carried through seed materials. Sugarcane setts are attacked by white ants, to avoid this problem, the setts are dipped in 50 percent BHC solution.

Seed-borne disease like loose smut of wheat can be controlled through drying for few hours in sun in summer months or hot water treatment for half an hour and then dried. To reduce a number of seed and soil-borne disease, the seeds are treated with fungicides like thiram, captan, mancozeb, carbendazim, captafol etc.

iii) Quickening or Hastening Germination

a) Deeping in water: In this process seeds are dipped in water for some time and then sown. This process is very effective in dry areas for early germination. Sometimes seeds are rubbed so that the hard and thick seed coat is separated, then it can absorb water quickly.

b) Scarification: Mechanically altering or softening the seed coats to make them permeable to water and gases is known as scarification. Mechanical scarification is generally done by pounding seeds in a large size mortar with pestle. Scarification is done at the time of sowing or a few days before sowing.

c) Hot water treatment: Seeds are dipped in water of 75 to 100°C. The volume of water is four to five times of seed. The seeds are soaked in gradually cooling water for 12 to 24 hours. Within a few days of hot water treatment seeds have to be planted.

d) Acid treatment: Dry seeds are placed in a container and concentrated sulphuric acid twice the volume of seeds are poured in it. The container should not be of metal or plastic. The treatment takes 10 minutes to 6 hours. The treatment period is completed when the seed coat is reduced. The extra acid is removed by washing the seeds in running water. Then the seeds are immediately sun or dried for storing.

e) Leaching: Leaching is done to remove the germination inhibitor through soaking the seeds in running water or placing in frequent change of water.

f) Hormonal treatment: To break the dormancy caused by physiological reason seeds are treated with gibberellic acid.

iv) Vernalisation

A very special kind of seed treatment in which seeds are treated to give rise to crops whose maturity is shorter than the normal periods. This is achieved through induced germination by soaking and incubating, then put in cold storage for varying period and from seeds are sown. The principle behind this that the plant spend a part of its vegetative phase in the sprouted condition, thereby period from sowing to flowering is reduced.

v) Inoculation with *Rhizobium*

Inoculation with *Rhizobium* helps the pulse crop to improve its nitrogen fixation capacity. When pulse crops are grown for the first time in any field, the seeds are treated with *Rhizobium* culture as soil may not have the particular *Rhizobium* species. The *Rhizobium* culture is mixed with jaggary solution and applied to the seeds and shade dried.

vi) Inducing of Variation

Seeds are treated for breaking type and inducing variation or mutation in the progeny. This is derived with the help of X-ray treatment.

Sowing

Before sowing following points should be considered:

i) Moisture content of soil for proper germination.
ii) Environmental condition, fair weather and assurance of rainfall condition are very important factor for sowing of seed.
iii) Crop variety which will determine how much spacing is necessary.

Time of Sowing

i) Sowing Early in the Season

Sowing early in the season may not be advantageous. For example, sowing rainfed groundnut in June may result in failure of the crop if there is prolonged dry spell from the second week of June to second week of July.

ii) Optimum Time of Sowing

Optimum time is the time of sowing at which all the growth stages of crop receive favourable environmental condition. Plant achieves huge flowering after sufficient vegetative growth. Moisture stress and other inconvenience may be avoided during critical stages.

iii) Delayed Sowing

Delayed sowing always reduces the yield. Reduction of yield is attributed to the early induction of flowering, unfavourable temperature and rainfall.

Methods of Sowing

Usually seeds are sown in three methods – (i) Broadcasting, (ii) Line sowing or drill sowing and (iii) Dibbling.

i) Broadcasting

Seeds are spread uniformly over well prepared land and is covered by ploughing or planking. It is the primitive and simple method. It is the most popular method among the farmers.

Advantages

a) No need of any special equipment.

b) Sowing can be done at wide range of moisture condition.

c) Less labour is required for sowing which reduces cost of cultivation.

Disadvantages

a) Uniform depth of sowing is not possible which affects germination.
b) Seeds are not covered properly and ultimately led to poor germination.
c) Requirement of seeds is more.
d) Later on intercultural operation is difficult.
e) Seeds are not placed at proper spacing.

ii) Line Sowing or Drill Sowing

Seeds are sown in uniform distant within the lines which are spaced at a definite distance. It is done by plough or seed drill at uniform depth.

a) Sowing Behind the Plough ('Kera' Method)

After final land preparation a plough man used to open the furrows at a regular spacing and another man used to drop the seeds in the furrow. After completion of this operation the seeds are covered by ladering.

b) Sowing with a Country Pora

Pora is conical cylinder made of bamboo splits. This pora is attached with the country plough at the back side of plough. After final land preparation, furrows are opened and seeds placed in pora drops at regular intervals through the narrow ends at the bottom which is adjusted by hand.

c) Sowing with Seed Drill

Sowing of seeds in line is achieved in a more controlled and regular manner by the introduction of seed drill.

Advantages of Line Sowing

a) Uniform depth of sowing is possible.

b) Seeds are covered properly.

c) Requirement of seed is low.

d) Later on intercultural operation becomes easier.

e) Seeds are placed at a proper spacing.

Disadvantages

a) Special equipment is required.

b) Sowing can not be done at wide range of moisture condition.

c) More labour requires for sowing which increase the cost of cultivation.

iii) Dibbling

This is adopted where distance from plant to plant and row to row is large and the seeds are large like maize. In this method, seeds are dibbled by hands.

Advantages

a) No room for mistake or uneven sowing.

b) Later operation becomes easier.

Disadvantages

a) Very slow process.

b) Cost of cultivation is more.

Depth of Sowing

Proper depth of sowing is essential prerequisite for good crop stand. Uneven depth of sowing results in uneven crop stand. Plants may be of different size and ages and ultimately harvesting will be problem as there is no uniformity in maturity. Shallow or deep sowing results in lesser plant population as all seeds do not germinate. There will be erratic plant population in the field. Weed problem become serious under this circumstances. It is therefore, essential to sow the crop at optimum depth for obtaining good stand of the crop. The optimum depth of bigger seed size like groundnut, castor, sunflower etc. can be sown even upto a depth of 6 cm. Small sized seeds like tobacco, ragi, ginger have to be sown as shallow as possible. If the seeds are sown to be shallow, the surface soil dries up quickly and germination may not occur due to lack of moisture. So, shallow sowing should be watered frequently ensure good germination.

The thumb rule for depth of sowing is 3 to 4 times to the diameter of the seeds to be sown. The optimum range between 3 cm to 5 cm. Shallow depth of 2 cm to 3 cm is adopted for small seeds like ginger, fingermillet and pearl millet. Very small seeds like tobacco are sown at 1 cm depth. This is generally done by broadcasting on the soil surface and mixing them by raking.

Seed Sowing

The production of high quality seed requires considerable technical skill; a number of rigid requirements must be fulfilled in order to ensure the high purity and germination requirements. Seed multiplication involves two separate steps (1) Seed Production and (2) Seed Processing.

Seed Production

Seed production, for convenience, may be discussed under two sub heads :

(1) Isolation and (2) Cultivation.

(1) Isolation

The crop raised for seed production should be separated from other fields of the same crop species by a minimum distance, which varies from one crop to the other. This distance is known as isolation distance. Isolation is essential for preventing pollination from unwanted pollen in the case of cross-pollinated and often cross-pollinated species and to avoid mechanical mixture and chance of cross-pollination in self-pollinated species. The isolation distance varies from 3 m in self-pollinated crops like, while, rice etc. while 200 m in the case of maize, bajra and jowar.

(2) Cultivation

Improved cultural practices and recommended levels of fertilizers and irrigation must be provided to raise healthy crop. The following points should be taken into consideration for raising good quality of seed.

Table: 1.1 Minimum distance for seed production of field crops

Crop	Contaminants				Remarks
	Field of other varieties		Fields of the same variety not conforming to varietal purity requirements for certification		
	Foundation	Certified	Foundation	Certified	
Paddy	3	3	3	3	1= Self compatible type, 2= Self incompatible types (in case of cotton/ jute, minimum isolation distance from fields of other species is 5 m both for foundation and certified seed production)
Paddy hybrids	200	100	200	100	
Wheat	3	3	3	3	
Maize	400	200	400	200	
Black gram	10	5	10	5	
Green gram	10	5	10	5	
Khesari	10	5	10	5	
Lentil	10	5	10	5	
Pea	10	5	10	5	
Bengal gram	10	5	10	5	
Pigeon pea	250	100	250	100	
Groundnut	3	3	3	3	
Linseed	50	25	50	25	
Sesame	100	50	100	50	
Soybean	3	3	3	3	
Sunflower	400	200	400	200	
Sunflower Hybrid	600	300	600	300	
Rapeseed and Mustard	200/1002	150/502	200/1002	150/502	
Potato	5	5	5	5	
Cotton/cotton hybrid	50	30	50	30	
Jute	50	30	50	30	

(a) **Land Requirement:** The land should be level, fertile and free from noxious weeds common to the crop. For some crops, like bajra and jowar, the field to be used for seed production should not have been used for growing the same crop in the previous year.

(b) **Cultural Practices:** Recommended dose of fertilizers and scheduled irrigation must be applied for obtaining good yield of quality seed. Poor cultural practices produce poor quality seed which will be discarded at the time of grading.

(c) **Plant Protection:** Insect pests and diseases may cause considerable damage to the crop reducing the yield and quality of seed. Incidence of some diseases leads to the rejection by certifying agencies. So, adequate measures must be taken against pest diseases.

(d) **Weed Control:** Weed not only reduces the yield of seed crop, but also weed seed contaminate the seed. So, special care must be taken against weed.

(e) **Roguing:** Roguing is the removal of off-type plant i.e. which are phenotypically different from the plants of the variety under certification. Off-type plants differ from desirable plants by height, leaf character, flowering time, maturity etc.

(f) **Special Operations:** Seed production in certain crops may require some special operations. For example, hybrid maize seed production may require detasseling of the female parent before they shed pollen.

(g) **Harvesting:** Harvesting and threshing of a seed crop should be done with considerable care in order to prevent mechanical mixture from other crop seed or weed seed.

Seed Processing

The seed obtained from threshing floor is contaminated with inert matter, other crop seed and weed seed (of different sizes), may have excess moisture and its germinability is not known. The various steps in seed processing are – (1) Drying, (2) Cleaning and grading, (3) Testing, and (4) Bagging and Labeling.

(1) Drying: Seeds are dried to an appropriate moisture level in order to facilitate processing, to prevent losses in germination and to reduce the chances of insect attacks during storage.

(2) Cleaning and Grading: Seeds from threshing floor contain several unwanted materials like weed seed, other crop seed, straw pieces, gravel, sand, soil particles etc. Seeds should be free from this material.

Seeds contain undersized, shriveled, broken seeds. Removal of these types of seed is known as grading.

(3) Testing: The seed lot is tested for percentage of pure seed, weed seed, seeds of other crops and germination after cleaning and drying. To check the quality of seed testing is essential for certification of seed. We shall discuss testing a little later.

(4) Bagging and Labeling: After seed treatment, seeds are distributed in bags of appropriate size, this process is known as bagging. Each bag is labeled with an appropriate label which carries the following information: (i) kind of seed, (ii) name of the variety, (iii) purity, (iv) germination percent, (v) date of germination test, (vi) weed seed percentage, (vii) inert matter percentage, (viii) name and address of the seller, (ix) period of variety of the certification and (x) any other information pertinent to the seed.

Genetic Principles of Seed Production

There are different reasons for arising variation among seeds –

(1) **Developmental Variation:** Seed production needs definite climatic condition, soil condition, altitude and other factors for maintaining quality.

(2) **Mechanical Mixture:** Seed production goes through the following steps : Sowing harvesting → processing → storage. During these steps seeds should not be mixed with other variety of the same crop.

(3) **Mutation:** There may be some off-type and abnormal plants. These should be rogued. Roguing is done at vegetative, flowering and maturity stage of plants.

(4) **Natural Crossing:** To check the natural crossing which creates variation can be checked through isolation distance.

(5) **Minor Genetic Variation:** At the time of release of the variety the plants are homogenous. But with the course of cultivation variation will occur among few plants. These plants are to be removed. Effect of this type of variation is less.

(6) **Selective Influence of Diseases:** The selective influence of diseases in varietal deterioration is also of considerable importance. New crop varieties often become susceptible to new races of diseases often caused by obligate parasites and are out of seed programmes. A more organized effort is required to check various diseases and produce seeds under disease-free conditions.

(7) **Techniques of Plant Breeder:** It takes 7-8 years to produce a variety. It needs screening and after screening of 7-8 years and when it is found that variety is fully purified, then it goes to the farmer. Premature release of varieties leads to deterioration of the variety.

Procedure for Maintaining Genetic Purity

(a) **Control of Some Source:** The use of seed of an appropriate class and from an approved source is necessary for raising a seed crop. Four classes of seeds, namely, breeder's, foundation, registered and certified are generally recognized in seed certification.

(b) **Preceding Crop Requirements:** These have been fixed to avoid contamination due to volunteer plants. While selecting a land for seed production it must be confirmed that same crop has not been cultivated in that lands in previous season.

(c) **Isolation:** The isolation of seed crops from various sources of contamination is also a necessary requirement for raising seed crop. Isolation is required during raising of a seed crop to avoid contamination due to natural crossing and disease infection, by wind and insects from neighbouring fields, and also during sowing, harvesting, threshing and handling of seeds to avoid mechanical mixtures.

(d) **Roguing:** The existence of off-type plants, i.e. plants differing in their characteristics from those of the seed variety is another potent source of genetic contamination. The removal of such plants is referred as 'roguing'.

(e) **Seed Certification:** The genetic purity in commercial seed production is often maintained through a system of seed certification.

(f) **Grow-out Tests:** Varieties being grown for seed production should periodically be tested for genetic purity by grow-out-tests, to make sure that they are being maintained in their true form.

Agronomic Principles for Seed Production

(1) Selection of Suitable Agro-climatic Zone

Dry climate, less humidity and bright sunshine is the best for seed production. High humid and high rainfall hampers pollination. Any abnormal climatic condition during flowering to seed production hampers seed production. High wind and cyclone in sea-coast does not permit seed production. During maturity of seed weather should be dry to facilitate drying of seed. Seed production of any crop should be done in a particular zone, *viz*. rice seed production should be done in Eastern India.

Table 1.2: Specific requirements for seed crop production

Crop	Maximum permitted (%)							
	Off-types		Inseparable other crop plants		Objectionable weed plants		Plants affected by seed borne disease	
	Foundation	Certified	Foundation	Certified	Foundation	Certified	Foundation	Certified
Paddy	0.050	0.20	-	-	0.010	0.020	-	-
Wheat	0.050	0.20	0.010	0.050	-	-	0.10	0.50
Maize	1.0	1.0	-	-	-	-	-	-
Black gram	0.10	0.20	-	-	-	-	-	-
Bengal gram	0.10	0.20				-	-	
Green gram	0.10	0.20	-	-	-	-	0.10	0.20
Khesari	0.10	0.20	-	-	-	-	-	-
Lentil	0.10	0.20	-	-	-	-	-	-
Peas	0.10	0.20	-	-	-	-	-	-
Pigeon pea	0.10	0.20	-	-	-	-	-	-
Groundnut	0.10	0.20	-	-	-	-	-	-
Rapeseed & Mustard	0.10	0.50	-	-	0.050	0.10	-	-
Soybean	0.10	0.50	-	-	-	-	-	-
Sunflower	0.10	0.20	-	-	-	-	0.050	0.50
Linseed	0.50	0.10	-	-	-	-	-	-
Sesame	0.10	0.20	-	-	-	-	0.50	1.0
Potato	0.050	0.10	-	-	-	-	-	-
Cotton	0.10	0.20	-	-	-	-	-	-
Jute	0.50	1.00	-	-	-	-	1.00	2.0

(2) Selection of Land

The land in which seed crop will be grown the cropping history of that land should be known. The land must be discarded if the land has been used to grow the same crop in the previous season or year. Otherwise mixture with previous year seed will be occurred.

The following points are to be considered during selection of land –

a) Soil fertility: It should be according to the crop requirement.

b) Texture: Texture of the land should be according to the crop need, *viz.* for paddy soil must be clayee.

(3) Isolation Distance of the Seed Crop

To avoid cross-pollination isolation distance must be maintained.

Crop	Isolation distance(m)
Paddy	3
Jowar	200
Maize	300
Pulse	10

(4) Preparation of the Land

For uniform germination of seed the land should be well prepared. Before sowing and is leveled properly and adequate drainage facility must be maintained.

(5) Selection of the Variety

Seed production of a certain variety must be followed in area where that variety has local popularity. The variety should be high yielding, posses disease resistance and other quality.

(6) Seed

It should be obtained from genuine source. The seed contains a tag (label) where the name of the crop, name of the variety and date of packing and expiring noted.

If we want to grow foundation seed, we must obtain breeder seed.

(7) Seed Treatments

Seed treatment before sowing the seed crop must be followed with the goal to prevent seed-borne disease and insect pests. Seed treatment also followed for inoculating nitrogen fixing bacteria.

(8) Time of Sowing / Planting

Normal time of planting or sowing must be maintained. In case of wheat, it is October to November.

(9) Seed Rate

Optimum seed rate must be maintained. In case of wheat 100 kg/ha is required generally. But in case of seed production, 90 kg/ha is required because here 100% seed germination is possible.

(10) Method of Sowing

The seed should be sown in line for better intercultural operations.

(11) Depth of Sowing

Different sowing depth is to be maintained for different seed size :

(i) For small seed, 3 - 4 cm deep

(ii) For medium seed, 5 - 6 cm deep

(iii) For large seed, 7 - 8 cm deep.

(12) Roguing

Roguing is a compulsory for any seed production programme. Roguing is done at vegetative stage, flowering stage and at maturity of the seed crop.

(13) Supplementary Pollination

This operation is needed for the cross-pollinated crops. Generally, pollination is done by any vector. If the production of vector is hampered then artificial method should be done. This operation is required for sunflower seed production.

(14) Intercultural Operation

Intercultural operations like weeding, thinning, earthing up should be done at proper time and in proper way. During intercultural operations weak plants must be removed.

(15) Disease – Pest Control

Disease and pest controlling chemicals must be sprayed in a routine work. Chemicals for controlling disease-pest may be used even before appearance of disease-pest under seed production programme.

(16) Nutrition of The Crop

Major plant nutrients like N, P and K are to be applied in proper dose. Some specific crop or area may require secondary nutrients and micro nutrients. For oil seed crop – N, P, K, S and B are compulsory.

In North Bengal soil, boron should be applied. Nitrogen is applied in 2 splits, ½ as basal and rest ½ as top dressing.

(17) Irrigation

Schedule irrigation must be followed.

(18) Harvesting

Before harvesting, the maturity of crop is judged according to some indication, like seed colour. The harvested produce should be kept and threshed specially to avoid the mixture with other varieties.

(19) Drying of Seed

Before storage, seed should be dried properly. If moisture content is high it will rot. The optimum moisture content is 9-11%. Germination and viability will be lost if moisture content is high.

For storage, seed bin should be used. Before storage of seed, it should be treated by some insecticides to control seed-borne pest.

1.2 Seed Testing

After cleaning and grading, the seed lots are tested for percentage of pure seed, weed seeds, seeds of other crops, inert matter and germination. This is known as seed testing. Seed tests consist of a series of tests designed to determine the quality of seed. Generally, the following tests are conducted to determine the quality of seeds : (1) Purity test, (2) Germination or seed viability test and (3) Moisture content test.

Sampling

Before seed testing, there is another process named 'sampling'. Seed tests are conducted on small samples generally drawn from processed seed lots. It is essential that samples used for seed tests are a representative of the lot.

Purity Test

Purity denotes the percentage of seeds (by weight) belonging to the variety under certification. The working sample is closely examined, often with the help of a magnifying glass, to classify it into the following components.

i) Pure seeds – Seeds of the variety under certification

ii) Seeds of the other varieties of the same crop.

iii) Seeds of the other crops

iv) Weed seeds

v) Inert matter – sand, straw, stones, pebbles, soil particles etc.

vi) Defective seeds – broken and shrunken seeds. A broken seed which is more than half of its original size and has intact embryo is classified as pure seed. The defective seeds are classified as inert matter.

$$\text{Purity}\left(\text{as percent}\right) = \frac{\text{Weight of pure seed}}{\text{Total weight of working sample}} \text{ x } 100$$

Weight of in pure seed: Weight of (pure seed + seeds of other varieties + seeds of other crops + weed seeds + inert matter)

$$\text{Similarly, Impurity } \left(\%\right) = \frac{\text{Weight of (seeds of other varieties + seeds of othercrops + weed seed + inert matter)}}{\text{Total weight of the sample}} \text{ x } 100$$

Impurity percentage is also referred to as dockage.

Germination Test

Germination is determined as percent of seeds that produce seedlings under a suitable environment. Thus germination test is of great importance because the sole function of seed is to produce healthy seedlings for raising a good crop.

It is the most important test to determine the quality of the seed. The seeds should germinate in the field to a satisfactory level, so that we can get optimum plant population in the field.

There are several process by which germination test can be carried out.

(1) Petridish Method

In this method a petridish of convenient size is used for testing the germination of the seed.

Process

i) First petridish is sterilized so that the microorganisms can not hamper.

ii) Blotting paper is put on it. Then it is moistened by distilled water.

iii) Put the seeds and count it.

iv) Blotting paper must be in wet condition, no stagnant water.

v) After 7-10 days, most of cereal and legume seeds will germinate completely.

vi) Germinated seeds are counted and removed regularly. The total number of germinated seeds would be the sum of the number of seeds that germinated at different observations.

$$\text{Viable seeds } (\%) = \frac{\text{Number of half seeds stained red}}{\text{Total number of half seeds}} \text{x } 100$$

(2) Rolled Towel Method

Two small wet towels are taken and seeds are put on it and roll it and roll it with another wet towel. Water is given as per requirement. After 7 days germination of seeds are observed.

(3) Sand Method

Aluminium tray of 23 cm x 23 cm x 8 cm is taken. Usually, sands are used as medium. About ¾ th of the depth is given and then water is poured in it. Then seeds are dipped in the sand, water is given periodically. After 7-8 days germination of seeds are seen.

(4) Seed Germination

It is small incubator where moisture and temperature can be controlled. There are many aluminium trays and a glass door in it. In trays filter papers are used as medium. On petridish seeds are placed and after 7 days germination result is observed.

Evaluation of Seedling

Normal seedlings should posses all the following structures –

i) A well developed root system including a primary root, except for those plants (e.g. Certain species of Graminae) normally producing seminal roots.

ii) A well developed and intact hypocotyls/or epicotyl without damage to conducting tissue and, in dycotyledons, a norml plumule.

iii) In the Graminae, a well-developed primary leaf within or emerging through the coleptile.

For Abnormal Seedlings

The seedlings which are badly broken, weak and malformed (deformed) are classified as abnormal seedlings. Abnormal seedlings are those which do not show the capacity for continued development into normal plants when grown in good quality of soil and under favourable conditions of water supply, temperature and light.

Seed Viability Test

Viability of seed can be defined as the capacity of the seed to remain capable of germination for some specific period of time. Sometimes it is necessary to procure a large quantity of seed within a very short time. Under that circumstances the germinability of the seeds can be tested only through the viability test. So, it is the quickest method through which we can know the germinability of the seeds, i.e. within 24 hours we can get the result. This test indicate whether the embryo is live or not. There are various methods through which this test can be performed. The most common method is Tetrazolium method.

Tetrazolium Method

This method determines the percentage of viable seeds which may be expected to germinate. The chemical 2, 3, 5-tetrazolium chloride or tetrazolium chloride in short, is colourless, but it develops intense red colour when it is reduced by living cells. This phemonenon is used to determine the percentage of viable seeds in a seed sample. Seeds are soaked in tap water overnight and are split longitudinally with the help of a scalpel so that a portion of the embryo is attached with each half of the seed. One half of each seed is placed in a petridish and covered with 1% aqueous solution of tetrazolium chloride for 4 hours. The seeds are then washed in tap water and the number of seeds in which the embryo is stained red is determined.

$$\text{Viable seeds } (\%) = \frac{\text{Number of half seeds stained red}}{\text{Total number of half seeds}} \text{ x } 100$$

Factors affecting viability –

i) Storage condition of the seed

ii) Age of the seed

iii) Environmental condition under which seed was developed

iv) The genetic make up of the seed

Moisture Content Test

Moisture content is determined as percent water content of seeds for satisfactory storage of the seed moisture content should be lowered down. Optimum moisture content reduces the deterioration during storage, prevents attack by moulds and insects and facilitates processing. The moisture content is determined by the drying the seeds in an oven or with the help of moisture metre.

In oven dry method, weighed seed samples are dried at 130°C for 90 minutes in an oven. The dried seeds are weighed again. The loss of weight represents the weight of water lost due to drying.

$$\text{Moisture content } (\%) = \frac{W_1 - W_2}{W_1} \times 100$$

Where, W_1 is the weight of seed sample before drying and W_2 is the weight of seed sample after drying.

Moisture metres measures the resistance of seeds to an electrical current; the electrical resistance of seeds varies with the moisture content. The use of moisture metres requires calibration and a certain degree of technical skill on the part of user. But moisture meters are very efficient and extremely rapid; a large number of samples can be handled in a relatively short period.

2

Seed Production of Cereal Crops

2.1 Rice

Rice is the staple food for the largest number of people on the Earth. According to an estimate, the global rice production must reach 800 million tonne from the present 91.04 million tonne in 2005-06 to meet the demand in 2025. It's area is 37% of the total area under food grains as against 20% under wheat, another important food crop. But globally it stands next to wheat in harvested area. As the irrigated rice contributes more than 75 per cent of the total rice production, enhancing its yield potential would be key to meeting the global rice requirement for an additional 215 million tonne. To achieve the target yield, rice varieties with a yield advantage of about 20 per cent over popular and widely adapted rice varieties under tropical conditions must be developed. Experience in China, India, Vietnam, Philippines, Bangladesh and several other countries clearly indicates that hybrid rice technology offers readily available option to meet this challenge. Hybrid rices have clearly shown a 1.0–1.5 t/ha yield advantage over modern high yielding varieties in the farmers fields in China and other countries.

2.1.1 Origin and Distribution

Rice is regarded as a first cultivated crop of Asia. Preserved rice grains were found in China around 3000 B.C. Rice grains found during excavation of Hastinapur (India) around 100-750 B.C. considered as a oldest sample in the world. Vavilov (1926) concluded that South-west Himalayas has various types and varieties and indicates probable centre of origin. De Condolle (1886) indicated South India as its center of origin. Thus based on all this Archaeological evidences India is considered centre of origin of Rice.

Rice cultivation is limited to as far as north as 49° in Czechoslovakia and as far as South as 35° in new South Wales. But most of Chi-world's rice is grown in tropics in which include countries of South and South-East Asia. West Africa and Central and South America. India stands first in area and second with regards to production, China is first.

2.1.2 Classification

Rice belongs to family graminae. There are 24 of species of genus *Oryza* but only two species are cultivation viz. *Oryza sativa* and *Oryza glaberrima. Oryza sativa* is most popular and grown in the most of the rice growing areas of the world, whereas *Oryza glaberrima* is mainly grown in South Africa. *Oryza sativa* is further grouped into three i.e. indica, japonica and javanica rice. The characteristics of each type are given below:

SI.No.	Characteristics	Indica	Japonica	Javanica
1.	Area of cultivation	India	Japan	Indonesia
2.	Plant height	Tall	Dwarf	Tall
3.	Leaves	Broad, light green, pubescent	Narrow, dark, green	Broad, hard. Light green
4.	Tillering	Very high	Medium	Low
5.	Grain	Thin, flattened elongated	Small and round	Broad and thick
6.	Shattering	High	Low	Low

2.1.3 Botany

The rice plant is an annual grass with round, hollow and jointed culms, flat leaves, fibrous root and terminal inflorescence called panicle. The panicle is a group of spikelets consist of two very small outer glumes. It is borne on the pedicel which connects withy the panicle branch. All parts above the 'outer glumes' are collectively clled the floret. It consists of the hard covering known as the lemma and palea (the glumes) and the complete flower is formed between them. The flower consists of six stamens and a pistil. The stamens composed of two called anthers borne on slender filaments, while the pistil consists of the ovary, style and stigma. The stigma is plumose is borne a style which in turn is an extension of the ovary. At the base of the flower are two transparent structures known as lodicules.

Two ecological factors, day length and temperature, mainly control the flowering of the rice plant. Temperature affects both the photoperiod-sensitive and photoperiod-insensitive varieties. Generally, high temperature accelerates and low temperature delays heading. The accelerated response to photoperiod caused by high temperature is reported to be an overall effect, not usually occurring at any specific stage of plant growth prior to flowering. Rice is considered to be short day plant. Nearly all varieties mature in a shorter time under a short photoperiod (around 10 hours) than under a large one (14 hours), but the degree of sensivity varies greatly among varieties. Varieties that are very sensitive to photoperiods may react very low intensities (1-100 lux).

Twilight may contribute significantly to the length of photoperiod under day length. In addition to panicle initiation, the developmental processes between initiation and anthesis may also be affected by the prevailing photoperiod (Vergara, 1967).

2.1.4 Climate

Rice is heat and water loving plant, requiring high temperature and adequate moisture supply. For germination a minimum of 10°C and for flowering a temperature of 22-25°C is considered as an optimum. Grain formation takes place at 20-21°C. Low temperature particularly during 10-12 days before heading causes considerable loss in grain yield because of sterile spikelets. At ripening temperature should be in between 20-25°C. Rice is a short day, thermosensitive plant but now many thermoinsensitive varieties have been developed. Rice is mostly grown in area receiving 1000-1500 mm rain. Crop needs heavy rain (125 mm) during vegetative phase and 200 mm monthly rain for low land and 100 mm for upland rice. In India it is grown from 80 ton 350 north latitude from sea level to mountainous regions upto 3000 meters.

2.1.5 High Yielding Varieties

2.1.5.1 Land Requirement

For seed production select fields in which rice crop was not taken in previous reason, unless the variety grown was the same and the field approved by the certification agency for varietal purity. In addition, the selected plots should be leveled and the soil preferably clay loam with a pH value 6.5.

2.1.5.2 Isolation Requirement

Rice is mainly a self-pollinated crop, though some cross-pollination occurs. The extent of natural cross-pollination varies from 0.1 to 0.4 percent. For pure seed production the seed field must be isolated at least by three metres from other fields of rice.

2.1.5.3 Cultural Practice

The rice crop can be grown by direct sowing or by transplanting. For seed production, it is desirable to grow rice under puddle and transplanting system. Growing of seed crop by transplanting method is described below:

2.1.5.4 Nursery

(i) Selection of Land for Nursery: Select land on which rice nursery; or rice crop, was not grown in the previous year. This is necessary to avoid varietal admixtures due to volunteer plants in the nursery itself.

(ii) Time of Sowing: Sowing of late maturating varieties in the nurseries should be done from 3rd week of May to 1st week of June and that of early duration varieties by 3rd week of June.

(iii) Preparation Nursery Bed: The soil be well pulverized by repeated ploughing. Then flood the nursery plot, puddle and leave it for two days to set with a thin layer of water. Prepare raised beds measuring 6 x 1.50 m with 0.5 m wide channel all around to facilitate drainage.

(iv) After Care: Obtain breeder's/foundation seed from a genuine source. Use 500 to 600 g seed per bed of 9 sq m of coarse variety and 400 to 500 g for fine variety. Apply 450 g of SSP per bed. The seeds may be broadcasted as such, or after sprouting in puddle beds.

After-care of nursery for three to four days, to prevent bird damage. It is necessary to keep the beds wet with the water all through. Drain-off excess water, if any, due to heavy rains. Keep the nursery free of weeds.

(v) Uprooting of Seedlings: Seedlings are ready for transplanting three to four weeks after sowing depending upon the variety. Generally for every month of crop duration seedlings remain in nursery for one week. Uproot the seedlings gently. Discard weak, diseased and the seedlings which are different in any way from the usual characteristic seedling of the variety.

2.1.5.5 Transplanting and Management of Crop

a) Land preparation is same as nursery bed. It needs 100 to 120 kg nitrogen, 50 to 60 kg phosphorus and 50 to 60 kg potash for one hectare of land. Full dose of phosphorus, ½ of the dose of nitrogen and potassium are applied as basal, ¼ of remaining N and K during mid tillering and other ¼ of remaining at panicle initiation.

b) 2-3 healthy seedlings are transplanted in a hill, for shy tillering variety 10 cm x 10 cm, for short and medium duration variety 20 cm x 15 cm and for long duration variety 20 cm x 20 cm or 30 cm x 20 cm spacing should be followed.

c) Allow 2.5 to 5 cm of water from transplanting to hard dough stage.

d) Weed control must be completed before heading. It may be done manually twice or thrice or may be achieved through spraying through chemicals like propanil @ 3 kg a.i./ha for grassy weeds and 2, 4-D @ 1 kg a.i./ha for controlling broad leaved weeds.

e) Plant protection schedule should be followed strictly for controlling pest-diseases. Among insect-pest stem borer, BPH, Gandhi bugs are major pests; these are controlled by spraying different chemicals like phosphamidon, endosulfan etc. Among the diseases blast, bacterial leaf blight and streak, brown spot, foot rot etc. are major. Bacterial diseases are controlled by spraying agrimycin and streptocyclin. Fungal diseases are controlled by spraying copper-oxychloride, Mancozeb and Benlate etc.

f) Roguing is done at vegetative, flowering and maturity stage.

2.1.5.6 Harvesting and Threshing

Harvesting is done when the crop is at right stage of maturity and moisture percentage is between 17 to 23 and preferably at 18% moisture.

2.1.5.7 Yield

Seed yield varies from 5.0 to 7.0 t/ha depending upon the varieties.

Basmati Rice

This rice is characterized by superfine grains with good aroma, excellent linear elongation of kernel on cooking which is almost double its length and soft texture of cooked rice. Good quality basmati rice have intermediate amylase content, low gelatinization temperature and medium gel consistency.

Varietal Improvement

Breeding efforts were started in mid-sixties for better quality of traditional basmati rice, but none of earlier evolved varieties could make a real impact and replace the popularly grown scented rice varieties like Pakistan basmati, Dehradun Basmati or Karnal Local. Breeding for high yielding scented rices was conducted at different Institutes like PAU, IARI, DRR etc. Emphasis was given to 4 important quality features i.e. aroma, superfine grain (length 6.5 mm), linear elongation on cooking and milking recovery.

Important Variety

IET 8580 (Kasturi) developed at DRR which was a cross between Basmati 370/CRR 88-17-1-4, the later parent being a derivation of Ratna/Domsiah, combined semi-dwarf stature, high yield potential with several desirable quality features of Basmati rice.

Pusa Basmati-1 released at IARI by developing a semi-dwarf scented culture IET 10364 (Pusa 615-140-10-1). Two varieties S-370 and Punjab-1 developed by PAU are accepted by the farmers. The another variety Haryana Basmati-1 is also popular to farmers.

Variety	Karnel length (mm)	Character L/B ratio	Karnel length on cooking (mm)	Elongation ratio (cooked)	Milling %	Head rice %
Pusa Basmati- 1	7.20	3.99	13.90	1.93	70.0	55.0
Kasturi	6.94	3.85	12.41	1.79	70.0	61.0
Haryana Basmati- 1	6.65	3.91	12.40	1.86	72.0	60.0
Ranbir Basmati	7.13	3.77	11.90	1.67	70.0	54.0

2.1.6 Agronomic Management

2.1.6.1 Sowing in Nursery

Sowing of basmati rice in nursery should be done in the first week of June at Western Uttar Pradesh, Delhi, Rajasthan and Jammu & Kashmir. In Punjab raising nursery during the last week of May ensures strong aroma and high yields. At Haryana, although late sowing has been practice in the past to ensure the quality, early June sowing is now followed to avoid the blast diseases. In non-conventional areas of Andhra Pradesh, Odisha, West Bengal, early planting in July proved ideal for higher yields and late planting proved better for obtaining good quality of rice.

2.1.6.2 Field and Crop Management

The field should be prepared in dry condition 3 weeks before transplanting and submerge the land in 5-10 cm of standing water. In situ green manuring with *Sesbania aculeata* or sunhemp can be done prior to transplanting of rice in the main field. Apply basal dose of fertilizers after final puddling and leveling the field. Maintain 2-5 cm of water throughout the growing season. Uniform leveling of land is very much effective for water management throughout the growing period.

2.1.6.3 Transplanting

Transplant 20-25 days old healthy seedling (2-3 nos/hill) in lines at a spacing of 20 cm x 20 cm for traditional basmati varieties and 20 cm x 15 cm for high yielding dwarf varieties. In case of delay planting, adopt shallow planting (2-3 cm deep), practice closer spacing of 15 cm x 15 cm for both the cultivars and increase the no. of seedlings (4-5 nos/hill).

2.1.6.4 Nutrient Management

Farm yard manure @ 5 t/ha should be added to the field before transplanting if green manure was not done. Apply basal dose of phosphorus (30 kg P_2O_5/ha), ½ of the nitrogen (30 kg N/ha), ½ of the potassium (15 kg K_2O) and 25 kg zinc sulphate/ha. Remaining ¼th of nitrogen (15 kg N/ha) and potassium (7.5 kg K_2O/ha) should be applied at maximum tillering stage and rest ¼th of both should be applied at panicle initiation stage. For high yielding dwarf basmati varieties the doses are 80-40-40 kg N, P_2O_5 and K_2O/ha + 25 kg zinc sulphate/ha. The method of application is same as traditional basmati rice. The field should be drained prior to top dressing with nitrogen and potassium.

2.1.6.5 Weed Control

Hand weeding twice at intervals of 3 weeks starting from 20-25 days after transplanting. Herbicide like butachlor or Thiobencarb @ 1.5 kg a.i./ha may be applied at 5-7 days after planting to control the weeds.

2.1.6.6 Plant Protection

Plant protection schedule should be followed strictly for controlling insect pest and diseases.

Disease

(a) Blast disease can be control by seed treatment with Carbendazim (Bavistin 50 WP @ 4 g/kg seed).

(b) Foliar application of Carbendazim 50 WP @ 0.1% (1 g/l) is recommended to control sheath blight.

(c) Apply Mancozeb (Dithane M-45) @ 25% during evening hours to control False smut and grain discolouration. Avoid spraying during flowering period.

(d) BLB disease can be control by reducing nitrogen application and apply in 2 or 3 splits along with potassium.

Insect Pest

(a) Seedling root of rice dip prior to planting in 0.02% chlorpyriphos for 12 hours or in 0.02% chlorpyriphos + 1% urea for 3 hours will be effective to control gall midge and stem borer upto 30 days.

(b) Soil application of Phorate or Carbofuran or Quinalphos @ 1.0 kg a.i./ha can control stem borer and gall midge in vegetative stage. At heading stage, two times foliar spray of chlorpyriphos or Phosphamidon @ 0.5 kg a.i./ha are more effective.

(c) Foliar spray with phosphamidon, monocrotophos or carbaryl @ 0.5 kg a.i./ ha are quite effective against green leaf hopper.

2.1.6.7 Roguing

For seed production rouging must be done at least at vegetative, flowering and maturity stage.

(a) Offtype plants should be rouged out once prior to flowering and then at flowering and maturity.
(b) The major roguing should be done just after flowering.
(c) The final roguing should be done when fields are near maturity.
(d) Rogue out wild rice plants, plants affected by stem borers and diseases (i.e. tungro virus and false smut).

2.1.6.8 Harvesting and Threshing

Mature seed should be harvested in time for getting good quality of seed. At this stage moisture content of rice seed varies from 17 to 23%. After harvesting the crops are threshed and winnowed to remove chaff, dust, empty husk and light grains. The seed must be dried to obtain 13% moisture content for proper storage.

2.1.6.9 Yield

Seed yield of basmati rice varies from 3.0 – 4.0 t/ha of different varieties.

2.1.7 Hybrid Rice

The term 'hybrid rice' refers to the first filial generation of cross between two genetically dissimilar parent plants. The commercial hybrid refers to a superior F1 which not only outyields its superior parents but also shows significant yield superiority by a margin of at least 1 t/ha over the best commercial variety of similar duration and grain quality.

Hybrid rice research in China started in 1964 and the first cytoplasmic male sterile line developed in 1972 from a male sterile plant found in a wild population in 1970. Consequently, the first commercial hybrid rice variety was developed in 1976 yielding 20 per cent higher than commercial inbred varieties of similar maturity duration. Since the release of this first hybrid, the area under hybrid rice has increased consistently in China. Research Programme on hybrid rice at the International Rice Research Institute started in 1978 with the initial failure of introduced Chinese hybrid to yield no better than inbred checks and the parental lines had poor grain quality and adaptability problems in tropics.

In India, the commercial exploitation aspect of heterosis largely remained neglected until the Indian Council of Agricultural Research established a goal – oriented network on hybrid rice research and development in 1989 with massive infrastructure development programme. These efforts got further boost in terms of training hybrid rice scientists and technologists abroad and procurement of sophisticated research equipment under the United Nations Development Programme on "Development of Hybrid Rice Technology for Large Scale Adoption" in India in 1991. The project has operated in two phases till 2004-05 together with ICAR component on "Development of rice hybrids under the Mission Mode Programme of National Agricultural Technology Programme (NATP) at 12 centers located in the potential hybrid rice growing areas of the country with the Directorate of Rice Research, Hyderabad acting as the coordinating centre.

The achievements in the development and commercialization of hybrid rice in the country are highly encouraging. Since the first release of hybrid APHR-1 by the Andhra Pradesh in 1994, extensive evaluations of hundreds of the test hybrids in the multilocation trials across the country followed by on-farm trials (OFT's) of the promising ones has resulted into commercial release of 15 hybrids in the country by the public sector. Additionally, seeds of 7 hybrids are being sold by the key private sector seed companies.

Since last one decade efforts on development of another type of male sterility sensitive to environment and derived by interaction of environment factors with nuclear genes have been developed in India for development of two-line hybrids. In the tropics, temperature controlled male sterility called TGMS is more practical to use. Research work to develop thermosensitive genetic male sterility system has been in progress at Pantnagar, Hyderabad and Coimbatore Centers of the country.

Our superfine aromatic basmati hybrid, Pusa Rice Hybrid-10 with an yield advantage of 1.2 to 1.6 tonnes/ha higher than the best inbred basmati variety is recently released for commercial cultivation.

The salient features of 22 rice hybrids released in India are given in Table 2.2.

Table 2.2 : Salient features of rice hybrids released in India

S.No.	Hybrid	Parentage	Releasing Agency	Year of release	Duration (days)	Yield in OFT (t/ha)		Yield advantage over check (%)	Area of adaptation
1	APHR-1	IR 58025A/Vanja IR 58025A/Vanjaran	SVRC	1994	130-135	7.14	5.27 (Chaitanya)	35.4	Andhra Pradesh
2	APHR-2	IR 62829a/mt u 9992	SVRC	1994	120-125	7.52	5.21 (Chaitanya)	44.2	Andhra Pradesh
3	MGR-1	IR 628929a/ir 10198-66-2R	SVRC	1994	110-115	6.08	5.23 (IR 50)	16.2	Tamil Nadu
4	KRH-1	IR 58025A/IR 9761-10-1R	SVRC	1995	120-125	6.02	4.58 (Chaitanya)	31.4	Karnataka
5	CNRH-3	IR 62829A/ Ajaya R	SVRC	1996	125-1 30	7.49	4.45 (Khitish)	37.4	West Bengal
6	DRRH-1	IR 58025A/IR 40750-82-2-2- 3R	SVRC	1996	125-130	7.30	5.50 (Tellahamsa)	32.7	Andhra Pradesh
7	KRH-2	IR 58025A/KMR-3	SVRC	1997	1 30-1 35	7.40	6.1 0 (Jaya)	31.8	Katnataka
8	PantSankar TJPRI 95-17A/ UPRI-92-133R Dan-1	UPRI 95-17A/UPRI-92-133R	SVRC	1998	1 15-120	6.80	5.30 (Pant Dhan-4)	28.3	Uttar Pradesh
9	CORH-2 IR 58025A/C20R	IR 58025A/C20R	SVRC	1998	120-125	6.25	5.20 (ADT 37)	20.2	Tamil Nadu
10	ADTRH-1 IR58025A/IR 66	IR58025A/IR 66	SVRC	1998	115-120	7.10	4.90 (ASD 18)	44.9	Tamil Nadu
11	Sahyadri	IR 58025A/BR 827-35-3-1-1-R	SVRC	1998	125-130	6.64	4.89 (Jaya)	35.8	Maharashtra

S.No.	Hybrid	Parentage	Releasing Agency	Year of release	Duration (days)	Yield in OFT (t/ha)		Yield advantage over check (%)	Area of adaptation
12	Narendra IR 58025 A/NDR 3026-3-1R SVRC 1998 Sankar Dhan 2	IR 58025 A/NDR 3026-3-1R	SVRC	1998	125-130	6.15	4.94 (Saryu 2)	24.5	Uttar Pradesh
13	PHB71	Not disclosed	CVRC	1997	130-135	7.86	6.14 (PR 106)	28.0	UP, Maharastra, Tamil Nadu and Haryana
14	Hybrid 6201	Not disclosed	CVRC	2000	120-125	6.18	5.03 (Jaya)	22.9	Eastern and some part of South India
15	Pusa PH 10	Pusa 6A/PRR-78	CVRC	2001	125-130	4.35	3.11 (Pusa Basmati 1)	39.9	Haryana, Delhi, Uttaranchal and western UP
16	PR 6444	6CO-2/6MO-5	CVRC	2001	135-140	6.11	4.91 (Jaya)	24.4	Part of South, Eastern and western India
17	RH 204	Not disclosed	CVRC	2001	120-125	6.89	5.62 (Local check)	22.7	South and North West India
18	PRH-122	Not disclosed	CVRC	2004	95-100	5.43	4.78 (Local check)	13.6	Uttaranchal, Punjab, Haryana, UP, Odisha, Bihar and Nagaland

S.No.	Hybrid	Parentage	Releasing Agency	Year of release	Duration (days)	Yield in OFT (t/ha)		Yield advantage over check (%)	Area of adaptation
19	HTA-126	Not disclosed	CVRC	2004	105-110	5.49	5.18 (Local check)	6.0	A. P. and Karnataka

Recently released varieties / hybrids of rice

Variety	Duration (days)	Yield (t/ha)	Area of adoption	Salient features
Bhupesh (IET 23324)	118-122	5.96	Tripura, Odisha, Bihar, Chhattisgarh, Rajasthan and Tamil Nadu	Suitable for all three seasons (*Pre-kharif, kharif* and *boro*), short bold grains, fertilizer responsive.
Telangana Vari-1 (IET 25330)	135	6.08	Odisha, West Bengal and Tamil Nadu	Suitable for irrigated conditions, moderately resistant to leaf and neck blast.
Bidhan Suruchi (IET 25701)	110–114	6.24 (*boro*); 5.6 (*kharif*)	West Bengal, Odisha, Uttar Pradesh and Assam	Suitable for both P-deficient and sufficient soils; Recommended for both *boro* and *kharif*, moderately resistant to bacterial leaf blight, leaf blast, stem borer and BPH.
Uttar Sona UBKVR-1 (IET 24171)	120	6.0 (*kharif*) ; 6.0 (*rabi*)	West Bengal, Bihar, Tripura and Assam	Suitable for both *boro* (irrigated) and *kharif* (irrigated and/or rainfed) early sown conditions, tolerant to drought at seeding stage. Resistant to LB and BS, stem borer and no incidence of BPH, WBPH, GM, LF and WM.
Uttar Lakshmi (IET 24173)	155–160	5.76	West Bengal, Odisha, Tripura	Suitable for irrigated (*boro*) in medium land situation, short bold grains, tolerant to stem borer, leaf folder, gall midge biotype 1 and whorl maggot.
MRP 5222 MEPH 126 (IET 25269)	145	6.05	West Bengal, Bihar, Odisha and Maharashtra	Suitable for irrigated rice ecology, moderately resistant to leaf blast, neck blast and bacterial leaf blight.

Variety	Duration (days)	Yield (t/ ha)	Area of adoption	Salient features
Manisha (IET 23770)	120–126	5.55	West Bengal	Suitable for irrigated lands, long slender grains, moderately tolerant to bacterial leaf blight (BLB), leaf blast and stem borer.
Chinsurah Nona-2 (Gosaba-6) (IET 21943)	135	5.2	West Bengal	Suitable for rainfed shallow lowland of coastal saline tract, short bold grains, tolerant to blast, moderately tolerant to sheath rot, leaf blast and neck blast.
CNRH-103 (IET 22866) (Hybrid)	110–115	5.3	West Bengal	Suitable for irrigated ecology, long slender grains, moderately tolerant to brown and leaf spot.
MRP 5626 MEPH-134 (IET 26198)	114	5.09	Bihar	Suitable for dry direct seeded-aerobic conditions with need based irrigation, resistant to leaf blast, neck blast, false smut; less damage due to gall midge biotype 4 and plant hoppers.
Sabour Harshit Dhan (IET 25342)	120-125	5.76	Bihar	Suitable for irrigated and medium upland conditions, moderately resistant to BLB, blast and brown spot and tolerant to stem borer, leaf folder and gundhi bug.
GNV 10-89 (IET 24716)	120–125	6.5-7.0	Karnataka	Suitable for normal, early and mid-late sown conditions, medium slender grains, moderately resistant to leaf blast.
GNR-6	100–105	4.0–4.5	Gujarat	Suitable for rainfed transplanted condition, long slender grains, moderately tolerant to stem borer, leaf folder and sheath mite.
GNR-7	110-115	5.74	South Gujarat	Short slender grains, moderately resistant to BLB, sheath rot, stem borer, leaf folder, sheath mite and tolerant to BPH.
GRH-2 (Hybrid)	130–135	6.0–6.5	Middle and South Gujarat	Moderately resistant to bacterial leaf blight, leaf blast, grain discoloration, sheath rot, brown plant hopper (BPH), white backed plant hopper (WBPH), leaf folder and stem borer.
Shalimar Rice-4	135–140	7.5–8.0	Lower belts of Kashmir valley (<1700 above MSL)	Medium bold grains, tolerant to blast.

Variety	Duration (days)	Yield (t/ha)	Area of adoption	Salient features
Shalimar Rice-5	135–140	4.5–5.0	Irrigated high altitudes of Kashmir valley (2,000–2,250 above MSL)	Short bold grains, tolerant to blast.
Kalanamak Kiran (IET 24753)	140–150	3.7	Irrigated medium and rainfed lowland areas of Uttar Pradesh	Medium slender grains, strong aroma, high zinc (20 ppm) content, moderately resistant to major diseases.
CSR 49 (IET 20329)	130–135	2.94	Uttar Pradesh	Suitable for irrigated, high fertility, transplanted and high alkaline conditions, moderately resistant to leaf and neck blast, brown spot and sheath rot.
CSR 52 (IET 23206)	125–135	2.88	Uttar Pradesh	Suitable for transplanted irrigated alkaline sodic areas, moderately resistant to leaf and neck blast, brown spot and sheath spot, tolerant to hoppers (brown and green), gall midge, case worm and whorl maggot.
LG 93.01 (Hybrid)	120–125	5.22	All ecologies of Uttar Pradesh	Moderately resistant to leaf and neck blast, brown spot, sheath rot and resistance against bacterial leaf blight.
Pant Dhan 28	128–130	5.66	Irrigated transplanted conditions in Uttarakhand plains	Moderately resistant to leaf blight and stem borer.
Pant Dhan 22	120–130	5.45	Irrigated transplanted conditions in Uttarakhand plains	Moderately resistant to leaf blight and stem borer.
JR 206 (IET 26079)	120–122	5.5–6.0	Rainfed and irrigated ecologies of Madhya Pradesh	Resistant to bunt and blast.
HKR 128	141	4.83	Irrigated paddy growing areas of Haryana	Suitable for timely transplanting, semi-dwarf, less incidence of WBPH, BPH and leaf folder.

Variety	Duration (days)	Yield (t/ ha)	Area of adoption	Salient features
Haryana Basmati-2	143	5.34	Irrigated paddy growing areas of Haryana	Resistant to stem borer, plant hopper, leaf folder, lodging and shattering.
BPT 2295 (IET 22188)	150–155	6.5-7.0	Irrigated lowland conditions of Andhra Pradesh	Suitable for *kharif* in place of BPT 5204 in Krishna zone, medium slender grains, resistant to BPH and BLB.
Pandu Ranga (IET 22611)	145	5.56	Entire saline coastal belt of Andhra Pradesh	Tolerant to leaf blast, brown spot and sheath rot diseases and stem borer. Salinity tolerant, deep root system.
Rice VGD 1	127–132	6.04	Tamil Nadu	Suitable for irrigated samba/late samba season, moderately resistant to blast and brown spot and leaf folder, aromatic variety for biryani making.
ADT 53 (IET 23955)	110–119	6.68	Tamil Nadu	Suitable for irrigated conditions, moderately resistant to blast, sheath rot, brown plant hopper, and leaf folder, brown rice contains high zinc (26.06 ppm) and iron (14.70 ppm).
MP 3030 (IET 25764)	120–125	6.91	Haryana, Uttarakhand, Maharashtra and Gujarat	Suitable for irrigated ecology, below average rain and rainfed conditions, moderately resistant to leaf blast and false smut.
Swarna Shakti Dhan (IET 25640)	115–120	4.96	Haryana, Odisha, Bihar, Jharkhand, Chhattisgarh, Gujarat and Maharashtra	Suitable for irrigated and rainfed, shallow under aerobic conditions, resistant to sheath blight and false smut, and tolerant to leaf blast, neck blast and brown spot.
Sandhya (IET 25508)	115–120	5.42	Chhattisgarh and Telangana	Suitable for irrigated and rainfed, shallow under aerobic condition, short slender grains, moderately resistant to leaf blast and brown spot.
US 380 (IET 25728)	121–125	5.62	Chhattisgarh and Madhya Pradesh	Suitable for irrigated and rainfed in aerobic conditions, resistant to leaf blast and neck blast and moderately resistant to brown spot and BLB.
US 303 (IET 25804)	130–135	5.4	Chhattisgarh, Maharashtra and Madhya Pradesh	Suitable for irrigated and rainfed conditions under high and medium fertility, resistant to leaf blast, RTD, sheath blight and brown spot and moderately resistant to glume discolouration and BLB.

Variety	Duration (days)	Yield (t/ ha)	Area of adoption	Salient features
CR Dhan 312 (IET 25997)	135–140	6.4	Maharashtra and Chhattisgarh	Suitable for irrigated conditions, moderately resistant to leaf blast, neck blast and tungro virus.
RH 9000 Plus (IET24931)	125–128	6.5	Maharashtra	Suitable for irrigated conditions under medium to high fertility soils during *kharif*, moderately resistant to BLB, leaf blast, neck blast, sheath rot, glume discolouration, tolerant to lodging with strong culm.
MRP 5433 MEPH-129 (IET 25793)	130	5.56	Maharashtra	Suitable for irrigated and transplanted conditions during *kharif*, moderately resistant to false smut, tolerant/resistant to brown plant hopper.
CAU-R4 (Eenotphou) (IET 22469)	140	3.8-4.5	Manipur	Suitable for semi deep water conditions having a water level of 50 cm, moderately tolerant to rice blast, rice brown spots, rice stem borer and rice gall midge.
CAU-R2 (Tomthinphou) (IET 23540)	125	4.2	Manipur	Suitable for rainfed direct seeded under real jhum rice agro-ecosystem of Manipur and NEH region, tolerant to rice blast, brown spots, rice gall midge and rice stem borer.
CAU-R3 (Mangalphou) (IET 22833)	95-120	4.0-4.5	Manipur	Suitable for irrigated medium and medium upland conditions, tolerant to blast, brown spot, stem borer, gall midge and resistant to tungro virus.
PR 127	140	7.54	Punjab	Suitable for low land irrigated rice ecosystem, resistant to bacterial blight.
Him Palam Dhan 1	120–125	2.1-3.2	Himachal Pradesh	Suitable for low and mid hills under upland conditions, resistant to blast especially in Himachal Pradesh conditions and tolerant to leaf folder and rice hispa.
Him Palam Dhan 2	120–125	5.57	Himachal Pradesh	Suitable for low and mid hills under irrigated ecology for transplanting conditions, highly resistant to leaf and neck blast and resistant to leaf folder and rice hispa.

2.1.7.1 Grain Quality and Production of Hybrid Variety

For large scale adoption of hybrids by farmers, it is essential that the enhanced yield should accompany good grain quality and ensure acceptability and profitability. Rice breeders studied this aspect intensively and concluded that hybridity per seed did not impair quality in terms of physical and chemical characteristics as long as both the parents possess acceptable grain quality. Therefore, the parents and hybrids are evaluated for key trials like intermediate amylase content, get consistency with white translucent grains prior to their release for commercial planning.

Among released hybrids DRRH-1, Narendra Sankar Dhan-2, ProAgro-6444 are blast resistance. Some of the hybrids tolerant to BLB resistance. Some of the hybrids tolerant to BLB (Sahyadri, Pant Sankar Dhan-1 and Pant Sankar Dhan-3, ProAgro-6444, PHB-71, NSD-2) and BPH (Proagro-6201, RH-204 and Pusa RH-10) are identified. KRH-2, Pant Sankar Dhan-1, PHB-71, CORH-2 and Proagro-6201 are performing better even under moisture stress condition with an yield advantage of more than 20% over the best check.

2.1.7.2 Hybridization

Rice was at one time regarded as anemophilous, but in 1898 Takahashi in Japan proved that it was mainly self-pollinated and succeeded in producing new varieties by crossing. In 1904, Kano in Japan started rice breeding by hybridization; as a result, by 1913 twenty new varieties selected through crossing were in farmer's hands. The work has been further developed in that country. (Matsuo, 1955) Concurrently investigations of a similar nature were conducted in Java and in 1913, van der stok produced a new variety by this method which possessed economic value (Copeland, 1924). Hybridization has since been used successfully by investigators in many countries.

Hybridization for the purpose of producing new and improved types involves a delicate technique. The anthers must be removed before they mature and pollen from another variety used to fertilize the stigma. This fertilized stigma must then be protected to prevent further cross fertilization from neighbouring plants until the flowering period has passed.

2.1.7.3 Technique of Hybridization

Many investigators have evolved their own technique of emasculation for rice hybridization. Various authors involve clipping the glumes and this necessitates germinating the seed produced on nutrient agar, while the hot water method suffers from the disadvantage that the straw is easily broken just below the panicle when it is bent over.

At present the two most commonly methods which are used are Ramiah's (1927) and a development of this evolved by Rhind (1950) and Rhind (1955)..

a) Ramiah's Method

In Ramiah's method a panicle which started blooming the previous day is selected. Spikelets already fertilized are removed and the panicle enclosed in a dark paper envelope, both the panicle and envelope being held in position by a stake driven into the mud near the plant. The enclosing in an envelope is done an hour or an hour and a half before the probable time of natural opening. The envelope is removed after about 10-15 minutes when it is found that all the spikelets which are to open that day have opened in a flush, the undehisced anthers hanging out. The anthers are immediately removed and pollination carried out when the pollen parent is ready. The spikelet close again but by the time pollen is ready reopening them is easy. After pollination, the rest of unopened spikelet are cut away and the panicle covered with a muslin bag for two or three days. It is desirable to allow the pollinated spikelets to develop in the open to attain their full development.

b) Rhind's Method

The method of forcing flowering developed by Rhind depends upon surrounding the panicles with warm saturated air and was described by Jordan (1957 b) who successfully employed the technique in Sierra Leone. Water at 43°C is carried to the field in a vacuum flask. Panicle in the third day of flowering is chosen for the female parent and all spikelets that have already flower are removed. The absence of stamens in these spikelets is readily seen when the panicle is viewed against the light. A little warm water is poured into an empty vacuum flask, shaken up and emptied. The panicle is then inserted in a flask which is closed by holding the fingers across the mouth. The flask is kept in this position for 1 to 3 minutes by which time stamens will protrude from several spikelets. The flask is removed and the anthers pulled from the open spikelets with forceps. It is found that a no. of spikelets open shortly after removal of the flask and if the panicle is held bend to one side the stamens of these all come out together and may be removed in once operation. The stamens are removed by their filaments but if only the anthers are approachable, a grip on these enables them to be removed without dehiscing. If any anthers are left inside the gumes, a gentle beating of the panicle with the hand will dislodge them and they can then be pulled-off. Any other unopened spikelet are then examined with a hand lens and any whose stigmas show signs of pollen are cut-off.

2.1.7.4 Hybrid Seed Production Technology

Hybrid seed production package has been optimized to yield 2.0 – 2.5 tonnes/ha of hybrid seed. It included synchronization optimization techniques using growth duration difference method with flag leaf cutting, GA_3 application (50 g/ha) at 5% panicle emergence and supplementary pollination at flowering for 5-7 days. Recently, agro-management technology for seed production of Pant Sankar Dhan 3 was refined at Pantnagar. Results for 2 years (2003-2005) have shown the highest yield of 4.3 t/ha for the hybrid (A/R) and 7.8 t/ha for the restorer male parent. The most promising results were from the technique involving 15 g/m^2 seed rate, 5 cm deep seeding, 150 kg N with three splits in the nursery and transplanting of 2 seedlings/hill with application of 210 kg N/ha in 4 splits in main field.

Following the package developed for hybrid seed production, average yield obtained in large scale seed production in the country are 2.0 – 2.5 t/ha. At the average procurement price of Rs. 30 to 40 per kg, the gross returns are Rs. 60,000 to Rs. 80,000 per hectare. The cost of cultivation is approx. Rs. 25,000 – Rs. 30,000 per hectare. Therefore, the net profit by hybrid seed production is Rs. 35,000 to Rs. 50,000 per hectare. This is very lucrative enterprise and hybrid seed growers in Andhra Pradesh, Karnataka and Maharashtra have greatly benefited by this activity. Hybrid seed production also generated additional employment for 60-80 persons/hectare, particularly the rural woman.

General techniques: Firstly seedlings are prepared for ‘R’ lines. Seeds of this plant types are broadcasted in seedbed, subsequent sowing of seeds of the same type are to be done at the 4th day of first sowing.

Now, seedlings of ‘A’ line are produced in a plot where sowing was done at the 9th day of first sowing of ‘R’ type in its fist plot. At the 25th day of sowing of ‘A’ line all the seedlings are uprooted and ready for transplanting. All the seedlings of ‘R’ line sown in two different dates are mixed together. Then all the seedlings of ‘A’ and ‘R’ line are transplanted in the main field following a specific sequential arrangement. Twelve rows of ‘A’ line and two rows of ‘R’ line plants are arranged in a spacing of 15x15 cm for ‘A’ line and 30x15 cm for ‘R’ line. At the time of panicle initiation, 1/2 or 2/3 prepared part of flag leaf is to be cut in order to avoid any pollination hazard. Next day of the cutting GA_3 is sprayed @ 40 ppm and second day GA_3 is also sprayed @ 60 ppm. Some mechanical measures may be taken to improve pollination. The grain obtained from ‘A’ type plant is to be considered as hybrid at F1 generation.

2.1.8 Constraints of Adaptation of Hybrid Rice Technology in India

The major constrains which are identified are –

i) Inadequate availability of pure seed of female parents and hybrids.
ii) Inconsistency in the performance of released hybrids
iii) Indiscriminate introduction of hybrids in non-recommended regions for cultivation
iv) High cost of hybrid seeds
v) Lack of action plan to cover the target area by different states in the country.
vi) Poor quality of grain of most hybrids.

2.1.9 Future Strategy

i) Enhancement of heterosis by development and use of improved CMS lines and restorers two line breeding, developing inter subspecific hybrids and bio-technological tools.
ii) Purification/improvement of parental lines utilizing conventional and molecular techniques.
iii) Improvement of grain quality and incorporation of resistance to some important diseases and pests.
iv) Refinement of hybrid seeds production technology for enhancing seed yield and making the technology cost effective.
v) Development of location and hybrid-specific package of practice for cultivation.
vi) Development of hybrids for *boro* season and shallow low land ecosystem.
vii) Reorientation of efforts for expediting seed production and technology transfer activities.

2.1.10 Conclusion

At end, it can be concluded that the hybrid rice technology has proven potential to meet any future demand of rice in the country. Quality seed production of potential hybrids is the real greatest challenge. Public/private partnership has to develop and emphasized in accelerating the techno logy adoption at much faster pace. Scientists need to look for further breakthroughs in yield potential utilizing the hybrid seed cost. Quality and resistance to stresses are getting combined into hybrids for greater stability through bio-technology means.

References

Aldas-Janaiah, Janaiah-A; Economic and political weekly. 2002, 37 : 42, 4319-4328.

Copeland, E. B. (1924) Rice. Macmillan, London.

De condelle (1886) Origin of Cultivated Plants, Kegan Paul, Trench and Co.London,

Fu-Ya Ping; Zhu-Zheng Ge; Xiao-Han; Hu-Guo Cheng; Si-Hua Min; Yu-YoungHong; Sun-Zong Xiu; Fu-YO; Zhu-ZG; Xiao-H; Hu-GC; Si-HM; Yu-YH, Sun-ZX, Chinese Journal of Rice Science. 2001, 15 : 2, 97-100.

Huang-Zai Jin, Shi-Guang Yue; Jia-Jun; Liu-Qing Xi; Chen-Shi Hong; Hu-He Liang, Hybrid Rice, 2004; 19(2) : 36-37.

Kalavathi-D; Ananthakalaiselvi-A; Vijaya-J, Seed Research. 2000, 28:1, 10-12.

Li JiMing; Yuan-Longping; Li-J. M.; Yuan-L P, Plant Breeding Reviews 2000, 17 : 15-158; 39 pp.

Ramiah, K. (1927). 'Artificial hybridization in rice', Agric J. India, 22, 17.

Ramiah, K. and Rao, M. B. V. N. (1953) Rice Breeding and Genetics. Ind. Council of Agric. Res. Sci. Monogr. No. 19.

Rhind, D. (1950) Notes on the Improvement of the Yields of Paddy in Ceylon. Govt. Printing Press, Colombo.

Rhind, D. (1955) 'Paddy notes, prepared in connection with the Mwea-Tabere (Kenya) Scheme' (unpublished).

Sridhar-Jarugula; Allah Rang; Bharaj-TS; Jarugula-S; Rang-A, Crop Improvement. 2002, 29:1, 24-27.

Virmani – SS; Mao-CX; Toledo-RS; Hossain-M; Jonaiah-A, Technical Bulletin Food and Fertilizer Technology Center. 2002, No. 156, 13 pp.

Vergara, B. S. (1967) Photoperiod and the flowering of rice varieties IRRI Saturday Seminar. Xu-Zheng Rong; Chen-DaiBin; Gao-Feng, Hybrid Rice. 2005; 20(4); 24-25.

Vaviloi, N.I. (1926). Studies on the origin of Cultivated Plants, Bulletin of Applied Botany and Plant breeding vol 16(2)

2.2 Wheat

Wheat (*Triticum* spp.) occupies the prime position among the food crops in the world regarding production. Wheat is the principal food of man in most of the area of the world. Wheat has become the most important source of carbohydrates in the majorities of countries in the temperate zones due to its excellent backing quality. Wheat is used in manufacture of bread stuffs, dextrose, alcohol and certain breakfast foods. The byproducts of wheat (i.e. wheat bran, standards middling, flour middling etc.) are higher in protein content than wheat itself and serve as a valuable protein supplements in various livestock rations. The durum wheats are processed in semolina from which macaroni and related products are made. Wheat straw may be used as feed for livestock, bedding materials on farms, manufacturing of papers etc.

2.2.1 Origin and Distribution

De Candolle believed that wheat originated in the valley of Eupharates and Tigris and spread from there to China, Egypt and other parts of the world.

Some American investigations believed that wheat probably originated in Syria and Palestine mainly on the evidence of Aaronshon, who brought a type of wild wheat to the USA from that area. Vavilov (1950) concluded that the origin of the durum wheat was probably in the region of Abyssinia, where as the whole group of soft wheat probably originated in the region of Western Pakistan, South-western Afghanistan and the Southern parts of mountainous Babohara.

Wheat (*Triticum* spp.) is grown under various climatic conditions between 47°S and 57°N latitudes and widely cultivated. A systemic culture of wheat culture on the globe includes the North American continent, Latin American including Southern Cone, Europe, West Asia and North Africa, South and East Africa, South Asia and Far East and Australia. Wheat cultivation in the India spans between 10°N in the Peninsular India and 35°N in the north-western Himalayan foothills. Tandon and Sethi (1987) included that wheat can be cultivated in North hill zone, Northern plains, North-western plain zones and North-eastern plain zones. Wheat is the main cereal crop of India after rice. The main wheat producing states are Uttar Pradesh, Punjab, Madhya Pradesh, Rajasthan, Bihar and Haryana.

2.2.2 Classification

Wheat belongs to the grass family Gramineae (Poaceae) and the genus *Triticum*. The species of *Triticum* can be classified into three distinct groups; (a) Diploids – Chromosome no. 2n=7 (b) Tetraploids – chromosome no. 2n=14, and (c) Hexaploids – chromosome no. 2n=21. Perceval (1921) described and recognized 18 species of wheat. The twelve species of *Triticum* are cultivated.

2.2.3 Botany

Wheat has both primary root system or temporary root system and secondary root system. Primary root appears at seedling stage till continues upto an age of 21-25 days. Thereafter, the permanent root system develops at crown root stage (21-25 days for Mexican wheat) and remains till maturity. The round or cylindrical stem along with hollow at internode bears leaf. Leaf has four parts leaf sheath, leaf blade, ligule and auricle. The flowering portion of wheat plant is called ‘ear’ or ‘head’ but botanically it is termed as spike. Spikelets are systematically arranged and are distributed along a common axis, borne on alternate sides of rachis, giving zigzag appearance.

2.2.4 Physiology of Flowering

Wheat is long day plant and it normally flowers in the long days of spring or early summer. Wheat favours the short days to increase the vegetative growth;

whereas long days hasten the formation of inflorescence. Spring wheat usually flower at any day length from less than eight hours up to continuous light, but the process is accelerated with an increase in day length. Winter wheats generally complete their life cycle most rapidly when given low temperatures during early stages of growth as well as long day and high temperatures during the later stage of growth. The reproductive phase commences with the tip of the growing point, differentiated into a spike. The spike is hold inside the last leaf, called flag leaf. The flag develops completely at this stage and obtain maximum leaf area of plant.

2.2.5 Hybridization

Traits such as increased number of fertilizer florets, length and density of spike, reduction in shattering, disease resistance and greater responsiveness to fertilizers without lodging have been genetically introduced by recent tools of mutation and hybridization. Release of semi-dwarf variety, 'Grain' in 1961 by crossing the Japanese cultivar Norin-10 with American winter wheat "Brevor" opened the era of modern wheat production. The international maize and wheat centre (CIMMYT) in Mexico is the major germplasm repository of the world, while the International Centre for Agriculture Research in Dry Areas (ICARDA), Syria collects and disseminates the 'durum' germplasm.

The tetraploid wheat probably developed through hybridization of a 7-chromosome wheat with a 7-chromosome related grass. The parent involved undoubtedly was cinkarn or its wild relative, while the wild grass species may have been *Aegilops speltoides* (Riley *et al.*, 1968). Emmer possibly was derived directly from wild emmer, although it too may have been a product of hybridization (Mangelsdorf, 1953). Hexaploids wheats appear to have evolved most recently. These wheats seem to have originated by hybridization of a 14-chromosome wheat with a wild 7-chromosome some relative of wheat, probably from the genus Aegilops (Maugelsdorf, 1953).

2.2.6 Climate

Wheat is cultivated over a wide range of latitudes from 60°N to 60°S and altitude of 3500 metres from the sea level in the tropics and subtropics. Wheat favours cool and moist weather during the vegetative growth period and warm and dry weather during grain formation. Growth and reproduction of the wheat plant is influenced by temperature, rainfall and light. The optimum temperatures for germination, vegetative growth and grain development period are 20-22°C (though it can germinate at 4°C), 16 to 22°C and 25°C respectively. Wheat crop thrive well in those areas where annual rainfall ranges between 25 to 180 cm.

2.2.7 Agronomic Management Practices

2.2.7.1 Land Requirements

Select a field in which wheat was not grown in the previous season unless the var grown was the same and approved by the certified agency for varietal purity. The field should be well drained and the soil neither too acidic nor too alkaline. In areas where zinc deficiency prevails that should be avoided.

2.2.7.2 Isolation requirement

Wheat is normally a self-pollinated crop but chances of natural pollination varies from 1-4%. It is sufficient to isolate seed fields with a strip of 3 m all around, which is planted with a non cereal crop or left uncropped. In cases where variety is susceptible to heavy infestation of loose smut a isolation of international standard requires 180 m whereas in Indian standard it is only 150 m.

2.2.7.3 Land preparation

Total 2-3 ploughing followed by laddering is sufficient to prepare a good field for sowing of wheat crop. Pre-sowing irrigation is must for uniform germination. Give a light shallow ploughing after pre-sowing irrigation and lavelled properly. Apply chlorpyriphos 3 ml/lit of water for controlling white ant and mole cricket.

2.2.7.4 Sowing parameters

For long duration variety, 1st fortnight of November and for short and mid duration variety 2nd fortnight of November is ideal for sowing. Obtain seed from a authentic sources and approved by certified agencies.

Seed crop are sown in rows either with seed drill or behind the plough or by hands in a depth of 5 cm.

Row to row distance should be made 20-22.5 cm to facilitate rogouing and inspection work.

75-85 kg/ha (For fine seed like kalyansona, @ 75 kg/ha and for bolder seed like UP 262 @85 kg/ha).

2.2.7.5 Nutrient recommendation

The recommended dose for wheat crop seed production is 80-120 kg N, 40-60 kg P_2O_5 and 40-60 kg K_2O/ha. Fertilizer should be applied on the basis of soil test report. The whole amount of P_2O_5 and K_2O and half of N should be applied as basal and rest half of N should be applied as basal and rest half of N should be applied during 1st irrigation or 20-25 days after sowing. Band placement of fertilizer N 5 cm away from the seed give better result. If zinc deficiency is noticed in the soil then Zinc sulphate @ 15-20 kg/ha may be applied at the sowing time.

2.2.7.6 Water management

Depending upon the soil type and the duration of the crop 4-6 irrigation should be applied for seed production of wheat. In each irrigation 5-7 cm depth of water should be applied. First irrigation should be given at CRI stage i.e. 20-25 DAS. Then maximum tillering stage 40-45 DAS then flowering stage 55-60 DAS and dough stage 70-75 DAS should be applied in our condition. Another two irrigations should be given in areas where the winter is prolong and that should be applied at tillering and milking stages.

2.2.7.7 Intercultural operation

Field should be kept weed free through repeated weeding for reducing the competition. The critical period for crop weed competition for wheat varies from 20-25 DAS. In areas where the infestation of *Phalaris minor* and *Avena fatua* the field should be treated with stale seed bed method.

Chemically weeds are controlled by spraying 2-4-D sodium salt @ 0.6 kg/ha. with 750 lit of water at post-emergence spray (30-35 DAS). For controlling *P. minor* and *A. fatua,* pre-emergence application of Tribunil @ 1.5 kg ai/ha with 750 lit of water should be applied. Besides, pre-emergence application of (2-3 DAS) Basalin (Fluchloralin) @ 1 kg/ha can be recommended.

2.2.7.8 Roguing

At least 2-3 roguing may be necessary to bring the seed plot to seed certification standard. First roguing should be done before flowering, 2nd after flowering and 3rd just before harvesting. Before flowering, roguing should be done on leaf and stem characters, after flowering roguing done on the basis of inflorescence or ear head characters and before harvesting roguing should be done on the basis of grain character, awn characters etc.

2.2.7.9 Harvesting and Threshing (Dead ripe stage)

Soon after maturity seed should be harvested to avoid shattering and to avoid uncertain weather condition. During the harvesting time the moisture content should be 14-16%. Immediately after harvesting, the cocks are bundles and taken to the well cleaned threshing floor for threshing. After proper sun drying, threshing should be done either mechanically or manually.

2.2.8 Drying, Cleaning and Bagging

To maintain good quality the seed must be cleaned and bagged immediately after drying at about 9% moisture content of the seeds.

The cleaned, nagged seed should be stored in a well protected godown with proper leveling.

2.2.9 Seed Yield

Average seed yield varies from 35-40 q/ha.

Table 2.3: Improved wheat variety currently recommended for cultivation under various cultural situations in different states.

States		Cultural condition	Varieties
Panjab, Haryana, Delhi, Western Utter Pradesh, Rajastan (except Kota & Udaipur Division), Foot-Hills of Uttaranchal, Himachal Pradesh, Jammu & Kashmir		All situations	WH 542, UP 2338, Raj 3077, PBW 343
	(a)	Irrigated timely sown.	HD 2687, CPAN 3004, PDW 215(d), PDW 233(d), WH 896(d), PBW 502
	(b)	Irrigated late sown	UP 2338, PPW 226, Raj 3765, Raj 3077, PBW 373 UP 2425,
	(c)	Irrigated very late	HD 2285, Raj 3777, Raj 3765
	(d)	Sown Rainfed	PBW 175, PBW 299, PBW 65, PBW 396, WH 533 in Hariyana, C 306
	(e)	Salt-affected soils	KRL 14, Raj 3077, PBW 65, KRL 19,
Bihar, Eastern Uttar Pradesh, West Bengal, Odisha, Assam, and Other North Eastern States.	(a)	Irrigated timely sown	K 8804 (k88), DL 784-3, HD 2402, HP1731 (Rajlaxmi), K9107(Deba), NW 1012, HUM 468, PBW 443, HD 2733, HD 2824,
	(b)	Irrigated late sown	HP 1633 (Sonali), HP 1209, HUW 234, HP 1744, HD 2643, NW 1014, HW 2045, NW 2036, DBW14.
	(c)	Irrigated very late sown	HD 2285, Raj 3777, Raj 3765.
	(d)	Rainfed timely soon	K 8027,
	(e)	Rainfed late sown	HDR 77, K 8962 (Indra), K 9465.
	(f)	Salt-affected soils	KRL 1-4, Raj 3077, PBW 65, KRL 19.

Madhya Pradesh, Gujarat, Kota & Udaipur Division of Rajasthan, Jhansi Division of Uttar Pradesh	(a)	Irrigated timely sown	DL 803-3, GW190, HI 1077, HI 8498(d), GW 273, HD 2236, Raj 1555(d), HI 8381(d), GW 1139(d), MP 4010
	(b)	Irrigated late sown	GW 173, Swati, J 405, DL 788-2, GW 322
	(c)	Irrigated very late sown	HD 2285, Raj 3777, Raj 3765
	(d)	Rainfed timely soon	C 306, Sujata, A 9-30-1 (d), HW 2004, JSW17, HD 4672(d), HI 1500
	(e)	Salt-affected soils	KRL 14, Raj 3077, PBW 65, KRL 19.
Karnataka, Maharashtra	(a)	Irrigated timely sown	MACS 2496, DWR162, DWR 1006(d)
			HD 2380, AKW1071 in Maharastra, HD 2781, Raj 4037
	(b)	Irrigated late sown	HD 2501, DWR 195, H1977, NIAW 34, HUW510 HD 2285, R 3777, Raj 3765,
	(c)	Irrigated late sown	NI5439, MACS 2846, N 59(d), MACS 1967(d) Bijaga yellow (d)
	(d)	Rainfed timely soon	K 9644, KRL1-4, Raj 3077, PBW 65, KRL 19.
	(e)	Salt-affected soils	NP 200, DDK 1001, DDK 1009
Uttarakhand, Himachal Pradesh, and Jammu & Kashmir	(a)	Mid altitude	HS 240,HD 2380, VL 738
	(a)	irrigated timely sown	VL 832
	(b)	Rainfed timely soon	HS 240,HD 2380, VL 738, VL 832
	(c)	Rainfed late soon	HS 207, HS 295, HS 420
	(d)	Rainfed early soon	VL 616, HS 277, VL 829
	(e)	High altitude	NPW42, HS365
	(f)	Very high altitude (Summer sown)	HS 375
Tamil Nadu (Nilgiri & Palni Hills)	i)	All situations	HW 971, HUM 318, HW1085, NUW 2044
	ii)	Dicoccum Wheat	NP 200, DDK 1001, DDK 1009

Source: Handbook of Agriculture, 2006, ICAR.

Recently released varieties / hybrids of wheat

Variety	Duration (days)	Yield (t/ha)	Area of adoption	Salient features
SHUATS-W 13 (AAI 13)	115–120	3.48	Uttar Pradesh	Suitable for timely sown, irrigated conditions, plant height 95–100 cm, resistant to brown rust and Karnal bunt.
P 12499 WH 1184	144	6.13	Haryana	Suitable for timely sown, irrigated conditions, plant height 99 cm (95–105 cm), highly resistant to yellow and brown rusts, protein content (13.02%), sedimentation value (60 ml).
Karan Narendra (DBW 222)	143	6.13	Punjab, Haryana, Delhi, Rajasthan (excluding Kota and Udaipur division), Western Uttar Pradesh (except Jhansi division), Jammu and Kathua district of Jammu and Kashmir, Paonta Valley and Una district of Himachal Pradesh and Tarai region of Uttarakhand	Suitable for timely sown, irrigated conditions, resistant to stripe and leaf rust.
Karan Shriya (DBW 252)	127	3.67	Eastern Uttar Pradesh, Bihar, Jharkhand, West Bengal (excluding hills), Odisha, Assam and plains of North Eastern states	Suitable for timely sown, restricted irrigation conditions and salt affected soils, highly resistant to rust and wheat blast.
Pusa Wheat 3249 (HD 3249)	122	4.88	Eastern Uttar Pradesh, Bihar, Jharkhand, West Bengal (excluding Hills), Odisha, Assam and plains of NE States	Suitable for timely sown, irrigated conditions, highly resistant to wheat blast and resistant to leaf and brown rusts. It is a bio-fortified variety with high zinc content (42.5 ppm).

Variety	Duration (days)	Yield (t/ha)	Area of adoption	Salient features
Pusa Wheat 3271 (HD 3271)	104 (NWPZ); 98 (NEPZ),	3.25	Punjab, Haryana, Delhi, Rajasthan (excluding Kota and Udaipur divisions), Western Uttar Pradesh (except Jhansi division), parts of Jammu and Kashmir (Kathua district), parts of Himachal Pradesh, (Una district and Paonta Valley), Uttarakhand (Tarai region), Eastern Uttar Pradesh, Bihar, Jharkhand, West Bengal (excluding hills), Odisha, Assam and plains of N.E. States	Suitable for late sown, irrigated conditions, highly resistant to leaf rust and stripe; moderately resistant to leaf blight, powdery mildew, Karnal bunt and flag smut.
Pusa Wheat 1621	147	5.04	Punjab, Haryana, Delhi, Rajasthan (excluding Kota and Udaipur divisions), Western Uttar Pradesh (except Jhansi division), Jammu and Kathua district of Jammu and Kashmir, Paonta Valley and Una district of Himachal Pradesh, Tarai region of Uttarakhand	Suitable for timely sown, restricted irrigation conditions, high levels of field resistant to stripe rust and field resistant to leaf rust.
Pusa Wheat 1628	102 (NWPZ); 94 (NEPZ)	3.28 (NEPZ); 3.7 (NWPZ)	Punjab, Haryana, Delhi, Rajasthan (excluding Kota and Udaipur divisions), Western Uttar Pradesh (except Jhansi division), Jammu and Kathua district of Jammu and Kashmir, Paonta Valley and Una district of Himachal Pradesh, Tarai region of Uttarakhand	Suitable for late sown conditions, high levels of field resistant to stripe and leaf rust, resistant to leaf blight, Karnal bunt and head scab.
PBW 771	120	5.03	Punjab, Haryana, Delhi, Rajasthan (excluding Kota and Udaipur divisions), Western Uttar Pradesh (except Jhansi division), Jammu and Kathua district of Jammu and Kashmir, Paonta Valley and Una district of Himachal Pradesh and Tarai region of Uttarakhand	Suitable for late sown irrigated conditions, high degree of resistant to yellow and brown rusts, high zinc concentration (>40 ppm). A MAS derived bio-fortified variety of wheat.

Variety	Duration (days)	Yield (t/ha)	Area of adoption	Salient features
Him Palam Gehun 2 (HPW 368)	176–185	2.6 (rainfed); 5.09 (irrigated)	Mid and low hills of Himachal Pradesh	Suitable for timely sown rainfed and irrigated conditions, high level of resistant to yellow rust and resistant to powdery mildew, leaf and stripe rusts.
DDW 47	121	3.73	Madhya Pradesh, Gujarat, Rajasthan and Chhattisgarh	Suitable for timely sown restricted irrigated conditions, protein content 12.69%, resistant to black and brown rust, high yellow pigment content (7.57 ppm).
Gujarat Wheat 1346	122	4.88	Maharashtra, Karnataka and plains of Tamil Nadu	Suitable for restricted irrigated timely sown conditions, protein content 12.36%, highly resistant to wheat blast, resistant reaction for leaf and brown rusts, Karnal bunt, powdery mildew, flag smut and leaf blight.
Pusa Wheat 8802	105–110	3.6	Maharashtra, Karnataka and plains of Tamil Nadu	Suitable for restricted irrigation and timely sown conditions, high level of field resistant to stem and leaf rusts and resistant to Karnal bunt, loose smut, flag smut and foot rot.
Pusa Wheat 8805	105–110	3.54	Maharashtra, Karnataka and plains of Tamil Nadu	Suitable for restricted irrigation and timely sown conditions, protein content 12.8%, high level of field resistant to stem and leaf rusts and resistant to Karnal bunt, loose smut, foot rot and better resistant to flag smut.
UAS-466	120	3.88	Madhya Pradesh, Gujarat, Rajasthan and Chhattisgarh	Suitable for timely sown, restricted irrigation conditions, resistant to leaf, stem rusts and resistant to loose smut.

Refrences

Mangelsdorf, P.C. (1953) Wheat Sci. Am. 189(1):50-59)

Pereival, J. (1921). the wheat plant. Genold Duckworth and Co.Ltd. Londaon

Riley, R.J. Unrau and chapman, V. (1958) Evidence on the origin of the B. genome of wheat J. Hered 49(3): 91-98.

Tandon, J.P. and Sethi, A.P. (1987). Twenty five years of co-ordination wheat research. Wheat Project Directorate. Indian Council of Agricultural Research, New Delhi, pp 270.

Vavilov, N.I. (1950). The origin, variation Immunity and Breeding Cultivated Plant. Chron. Bot. 13(1-6) (Trans by K. Starr Chester) 1949-50.

2.3 Maize

Maize is used directly as food at sight of the production. It is used locally as the major cereal in the traditional areas by the local urban population in the form of chapatti. In addition, green ears are consumed roasted form. Bakery production has played an important role in the human diet in the urban areas. Use of breed buns and biscuits etc. is becoming popular. Maize starch is a polymer of anhydrous glucose units (dextrose) and is the raw material for sweeteners. Many industries have come to derive starch from maize. The maize is also used for making maize syrup, high fructose corn syrup (HFCS), maize oil for cooking, alcohol and organic/synthetic co-polymers.

Maize seed is produced in two ways – (I) by Open Pollination and (II) by Hybridization.

2.3.1 Origin and Distribution

Maize has been cultivated for thousand of years and is generally agreed that the maize (*Zea mays* L.) plant is a native of the Americans. various authors (1955) indicated that pollen grains of *Zea tripsacum* and *Zea ecuchaena* are said to be found at depth of about 50 meters under Maxico city. At the time of discovery of new world, maize was already cultivated in many parts of the American continent, especially in the more highly developed region of Maxico, Central America and South America where it formed the staple diet of the indigenous population. Christopher Columbus writes of the new Graminaceae found in the under region of Cuba described by two members of this crew. Maize spread over large area of Spain, Portugal, France, Italy, South-eastern Europe and North African the years following its introduction. In terms of world acreage, India stands only next to the USA, Brazil, China and Maxico, where as with regards to production it ranks eleventh and mention here in order of production in USA, China, Brazil, Argentina, Maxico, France, South Africa, The former USSR, Rumania, Yugoslavia and India.

2.3.2 Classification

The genus *Zea* has only one species *Zea mays*, is so rich varieties that it present a very difficult task to botanist. So far as a unitary classification concerned, botanist (1895) classifies the maize in the following categories:

1. Dent corn - *Zea mays indentata*
2. Flint corn - *Zea mays indurate*
3. Sweet corn - *Zea mays saccharata*
4. Floury corn - *Zea mays amylacea*
5. Pop corn - *Zea mays averta*
6. Pod corn - *Zea mays tunicate*
7. Waxy corn - *Zea mays ceretina*

The several thousand varieties are divided into various groups according to peculiarities, such as the structure of plant. Well known systems are those which has been given by Koernicke (1873), Von Sturlevant (1899), Kuleshow (1933) and Grepenseikov (1949).

2.3.3 Botany

The maize stem is generally 2-3 meters high. The stem generally attains a thickness of 3-4 cm. On an average a plant posses fourteen internodes though at the apex of the plant, became longer and thicker higher up the stem, and then taper again to the male inflorescence which terminate the axis. The terminal root which develops from the radich with its lateral roots may persist throughout the life of the plant. The main adventitious fibrous roots developing from the lower nodes of the stem below ground level are the effective and active roots of the plant. The male and female inflorescence called tassel and cob respectively. The maize grains are fully exposed naked and more or less oblong. On the flat face of the grain, the embryo can be seen as a whitish dettoid area. The colour of the grain may be yellow or reddish or while and in some cases even brown and black. The upper end of the grain is often rounded or flattened.

Two types of inflorescence (1) Female inflorescence develop an ear (cob), (2) Male flowers (tassel).

The male flowers are borne in a cluster (tassel) on the top of the stem end as a terminal panicle, while the female flowers are borne inside the young cobs which spring from one of the nodes on the stem usually located about midway on the stalk.

At anthesis, just prior to shedding of pollen, the lodicules swell to several times than normal size and push the palea and lemma apart, making it possible for the anthers to be extruded by the elongating filaments. Anthers then break open near the tip, forming pores through which the pollen are shed in huge number in the wind.

2.3.4. Open Pollination

2.3.4.1 Land requirement

Maize can be grown in any type of soil provided there is no water-logging. It must be ensured that in the selected plot in previous season maize has not been cultivated.

2.3.4.2 Isolation distance

Isolation distance for foundation seed 400 m and for certified seed it is 200 m.

2.3.4.3 Cultural requirements

2.3.4.4 Land preparation

5-6 ploughings followed by laddering is sufficient for land preparation.

2.3.4.5 Sowing parameters

i) *Kharif* season – Mid June to Mid July

ii) *Rabi* season – October to November.

Row to row distance 60-75 cm and plant to plant distance 20-25 cm. Depth of sowing is 5-6 cm. Seed rate is 20 kg/ha.

2.3.4.6 Fertilizer application

For seed production maize needs 120-150 kg N/ha, 60 kg P_2O_5/ha and 40 kg K_2O/ha. Full dose of P_2O_5 and K_2O as basal, 1/3rd of N as basal, 1/3rd at one month after sowing and rest 1/3rd N at flowering should be applied. Sometimes 25 kg $ZnSO_4$/ha is needed in the area of zinc deficiency.

2.3.4.7 Water management

In summer, maize needs 5-8 irrigation, need less irrigation depending on rainfall. Maize is sensitive to water logging, so there must be drainage system.

2.3.4.8 Interculture

Maize field must be maintained weed free. It needs two interculture operations – (i) within three weeks of sowing and (ii) 15 days after 1st weeding. At the time of second weeding earthing up is done.

2.3.4.9 Pest diseases

Main insect-pests are stem borer and hairy-caterpillar. Hairy caterpillar is controlled by Endosulfan 35EC @ 1.5 ml/litre of water.

2.3.4.10 Roguing

All the off type plants are to be removed. During seedling stage 1st roguing is done and it must be completed before pollen-shedding.

2.3.4.11 Harvesting

After harvest cobs must be dried to reduce the moisture content and seeds are separated from the cobs. Again seeds are dried, for storage moisture content should be below 12%. Seed yield 2.5-3.0 tonnes/ha.

2.3.5 Hybrid Maize

2.3.5.1 Maintenance of inbred parent lines

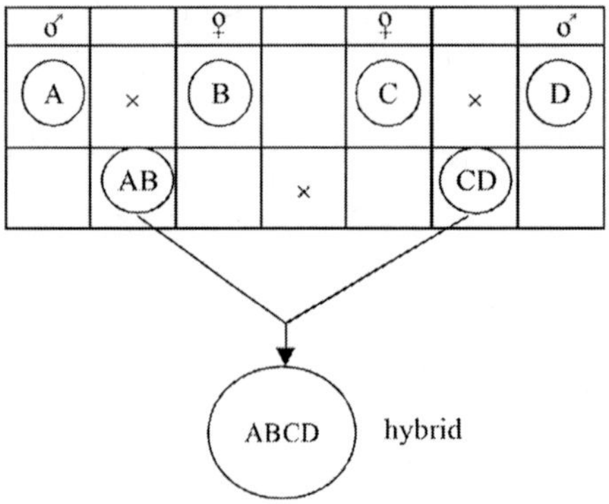

(A) is taselled and (B) detasselled.

2.3.5.2 Isolation distance : 400 m

2.3.5.3 Cultural requirement : as open pollinated crop.

2.3.5.4 Roguing

Inbred lines should be properly maintained. Any off-type should be rogued out before pollen shedding. First roguing is done when the plants are about knee height (1.5 ft). Second roguing is done just before opening of male flower. Final roguing is done after seed setting at this time diseased plants are also removed.

2.3.5.5 Seed yield : About 0.6 – 0.8 tonnes/hectare.

2.3.5.6 Precaution

a) During detasselling, all the male flowers are to be carefully removed. This operation should be done in the field everyday until the flowering is completed.

b) Some times tillers also come from the maize plants. So the tillers are to be removed.

c) During handling of the seeds care should be taken so that there is less chance of mixture with other varieties.

Table 2.4: Important cultivars of maize grown in India

State	Types of cultivar	Variety
Uttar Pradesh	Hybrid	Ganga 11, KH 528, PRO 311, Bio 9681, PRO 303, Deccan 107, Deccan 105, KH 945, MMH133, Sartaj
	Composite	Prabaht, Navjot, Pusa Composite 2, Arun, Kiran, Megha, Mahi Kanchan, Dhawal, Diara 3, Navin Azad, Uttam
West Bengal	Hybrid	Ganga 11, Bio 9681, Deccan 103, Deccan 105
	Composite	Prabaht, Navjot, Pusa Composite 2, Diara 3, Mahi Kanchan, Dhawal, Suwan 1
Rajasthan	Hybrid	Ganga 11, Ganga 2, KH 528, PRO 311, Deccan 107, Deccan 109, Deccan 105, KH 510, Pusa Early Hybrid Makka 1, Pusa Early Hybrid Makka 2, Him 129,
	Composite	Prabaht, Navjot, Pusa Composite 2, Arun, Kiran, Mahi Kanchan, Dhawal
Madhya Pradesh	Hybrid	Ganga 11, PRO 311, Deccan 107, KH 510, Deccan 103, Deccan 105, Him 129
	Composite	Prabaht, Arun, Kiran, Mahi Kanchan, Dhawal, Pusa Composite 1, Pusa Composite 2
Gujarat	Hybrid	Ganga 11, KH 528, Pusa Hybrid 1, Pusa Hybrid 2, Him 129, Deccan 103, Deccan 105, Trishulata
	Composite	Prabhat, Arun, Kiran, Mahi Kanchan, Pusa Composite 1, Gujrat Makki 1, Dhawal
Maharastra	Hybrid	Ganga 11, Ganga 2, Deccan 103, Deccan 107, PEHM1, PEHM 2
	Composite	Prabhat, Navjot, Kiran, Mahi Kanchan, Dhawal, Panchaganga
Punjab	Hybrid	Ganga 11, PRO 311, Paras, Deccan 107, Deccan 109, Prakash
	Composite	Prabhat, Navjot, Kiran, Mahi Kanchan, Dhawal, Pratap, Megha, Pusa Composite 1, Pusa Composite 2
Delhi, Haryana	Hybrid	Ganga 2, Sartaj, Paras, PRO 303, Deccan 107, Deccan 109
	Composite	Prabhat, Navjot, Kiran, Mahi Kanchan, Dhawal, Pusa Composte 2
Southern India	Hybrid	Deccan 103, Ganga 11, Deccan 107, PEHM 1, PEHM 2, KH510, PRO 311
	Composite	Prabhat, Navjot, Kiran, Harsha, Mahi Kanchan, Dhawal, Renuka

Recently released varieties / hybrids of maize

Variety	Duration (days)	Yield (t/ha)	Area of adoption	Salient features
Shalimar QPMH-1	135	5.88 (station); 5.57 (AICRP trials)	Jammu & Kashmir	Suitable for cultivation in *kharif* season, especially in low to medium elevation of Kashmir valley, Tryptohan content: 0.81%, Lysine content: 2.95%, protein: 10.71%. It has shown resistance to *Turcicum* leaf blight and common rust diseases.
Shalimar Sweet Corn-1	105–110	10.6 (station); 6.24 (AICRP trials)	Jammu & Kashmir	Suitable for rainfed, *kharif*, hilly regions, it has 14% TSS and shown resistance to *Turcicum* leaf blight and common rust.
SHIATS Makka-2	85–90	3.5–4.0	Uttar Pradesh	Suitable for *kharif* season, composite white seeded, resistant to maydis leaf blight, moderately resistant to *Turcicum* leaf blight and to stem borer.
PMH 12	90	7.11	Bihar, West Bangal, Jharkhand, Parts of Uttar Pradesh and Odisha	Suitable for irrigated conditions during *kharif* season, moderately tolerant to maydis leaf blight, *Turcicum* leaf blight, banded stripe downey mildew and post flowering stalk rot.
VL Sweet Corn Hybrid 2	100	10.35	Jammu & Kashmir, Himachal Pradesh, Uttararkhand NEH	Suitable for rainfed agro-ecosystem in *kharif*, moderately tolerant to maydis leaf blight, *Turcicum* leaf blight, Sheath blight, *Curvularia* leaf spot.
Maize VGIH 1	100	6.94 (low plant density 66,000- 6.49 (high plant density 83,000- 13.5	Bihar, Jharkhand, Odisha, Uttar Pradesh and West Bengal	Suitable for *kharif* season, resistant to *Curvularia* leaf spot and Rajasthan downy mildew and moderately resistant to maydis leaf blight, *Turcicum* leaf blight, banded leaf and sheath blight, charcoal rot, *Fusarium* stalk rot and bacterial stalk rot, possesses good quality attributes.
JKMH 4152	94–96	8.23	Punjab, Haryana, Delhi, Uttarakhand, Assam and Uttar Pradesh	Suitable for irrigated condition of rainy season (*kharif*), resistant to *Maydis* leaf blight and charcoal rot, moderately resistant to *Chilo partellus*. Non-shattering.

Variety	Duration (days)	Yield (t/ha)	Area of adoption	Salient features
PAC 751	100–105	7.16	Madhya Pradesh	Suitable for irrigated and rainfed ecosystem, resistant to common rust, brown stripe, downey mildew, moderately resistant/ tolerant to *Turcicum* leaf blight, *Maydis* leaf blight, *Polysora* rust and *Chilo partellus*.
PAC 753	100–105	6.82	Madhya Pradesh	Suitable for irrigated and rainfed ecosystem, resistant to common rust, Rajasthan downey mildew and *Curvularia* leaf spot, moderately resistant/ tolerant to *Turcicum* leaf blight, *Maydis* leaf blight and *Chilo partellus*.
Pusa HQPM-5	100	7.26 (NHZ); 7.51(NWPZ); 5.35 (NEPZ); 7.12(PZ); 5.12 (CWZ)	NHZ: Jammu & Kashmir, Himachal Pradesh, Uttarakhand (Hill region), North Eastern Hill Region (Meghalaya, Sikkim, Assam, Tripura, Nagaland, Manipur, Arunachal Pradesh) NWPZ: Punjab, Haryana, Delhi, Uttarakhand, (Plain), Uttar Pradesh (Western region) NEPZ: Bihar, Jharkhand, Odisha, Uttar Pradesh, (Eastern region), West Bengal PZ: Maharastra, Karnataka, Andhra Pradesh, Telangana, Tamil Nadu CWZ: Gujarat, Madhya Pradesh, Chattisgarh, Rajasthan	Suitable for irrigated conditions under *kharif* season, late in PZ, MAS derived hybrid, introgression of crtRB1 and lcyE gene, high Provitamin A (6.77 microgram/g) (1.02 in normal maize), lysine 4.25% and tryptophan 0.94% (Normal maize <0.6% tryptophan, <2.5% lysine).
Pusa Vivek Hybrid-27	90	4.85	Bihar, Jharkhand, Odisha, Uttar Pradesh (Eastern region), West Bengal	Suitable for irrigated conditions under *kharif* season, MAS derived hybrid, introgression of crtRB1 gene, improved Provitamin A (5.49 μg/g).
Pusa HQPM-7	90–95	7.45	Maharashtra, Karnataka, Andhra Pradesh, Telangana, Tamil Nadu	Suitable for irrigated conditions under *kharif* season, MAS derived hybrid, introgression of crtRB1 and lcyE gen, high provitamin A (7.10 microgram/g), lysine (4.19%) and tryptophan (0.93%).

Variety	Duration (days)	Yield (t/ha)	Area of adoption	Salient features
Pusa Super Sweet Corn 2	110	13.2 (NHZ), 13.6 (NWPZ), 11.8 (NEPZ), 13.8 (PZ), 11.4 (CWZ)	Himachal Pradesh, Haryana, Uttarakhand, Uttar Pradesh, Tamil Nadu, Karnataka, Chhattisgarh, Rajasthan	Suitable for irrigated conditions, high fertility soils during *kharif* season, average cob yield 10.0 t/ha (NHZ), 9.5 t/ha (NWPZ), 8.1 t/ha (NEPZ), 10.2 t/ha (PZ), higher sweetness (15.2% to 17.1 % brix), 50% silking in 53 days, moderately to highly resistant to major diseases and pests.
Pant Popcorn 1	87–92	3.98 (NHZ); 5.0 (PZ)	Jammu & Kashmir, Himachal Pradesh, Uttarakhand (Hills), Assam, Arunachal Pradesh, Manipur, Meghalaya, Mizoram, Nagaland, Sikkim, Tripura, (North Eastern Hill Region), Tamil Nadu, Karnataka, Andhra Pradesh, Telangana, Maharashtra	Suitable for irrigated conditions of *kharif* season, moderately tolerant to *Chilo partellus* under field conditions.
PMH 11	96	5.74	Punjab	Suitable for irrigated conditions during *kharif* season, moderately resistant to charcoal rot, *Maydis* leaf blight, bacterial stalk rot, banded leaf and sheath blight, and stem borer.

Maize used in different purpose

Fodder	: African tall, J 1006
Sweet corn	: Madhuri, Priya
Baby corn	: Amber pop, VL Amber pop, Pearl pop
Pop corn	: VL 42, Prakash, Parvati

3

Seed Production of Pulses

3.1 Redgram or Pigeon Pea

In any production programme, the importance of quality seed needs hardly be emphasized. In fact, the major bottleneck in the pigeon pea cultivation programme is the restricted or poor availability of quality seeds of the desirable varieties. The seed has been considered as the 'vehicle of change'. With better attention paid to their interaction among water, fertilizer and seed of improved varieties, the production can be maximized. Although a numbers of improved varieties of pigeon pea have been released since 1970, organized seed production programmes have been rather limited. Most of our national efforts in this direction continue to be centralized on cereal crops like wheat, rice and hybrid millets. Moreover, the benefits of the current research will not reach the farmers unless large-scale seed multiplication programmes are organized. Most farmers continue to use seeds of traditional varieties and consequently have not been able to expedite the higher production potential of the new strains. Now, the programme of seed production requires strengthening and its subsequent development to a stage where participation of private enterprises in seed business becomes a practical and profitable proposition. The Government can also support the private and voluntary agencies in the production and supply of improved seeds.

3.1.1 Origin and Distribution

Pigeon pea is an essential African grain legume, which has been cultivated in the Nile Valley for more than 4,400 years. It is also a very old crop of our country and some scientists still believe that it has been originated in India. But the crop might have been introduced into India by ancient traders trading on the route between Zanzibar, India and Ceylon.

Pigeon pea is commonly grown in several tropical and sub-tropical countries of Asia and Africa. It is one of the most prized pulse crops of India. Pigeon pea occupies an area of about 3.61 million ha with a total production of 2.7 million ton and an average productivity of 747 kg/ha. Pigeon pea is grown throughout the arid to dry land areas of the country. It can be grown in irrigated areas also. The predominant states are Maharashtra (highest area and production), Haryana (highest productivity), Uttar Pradesh, Punjab, Madhya Pradesh, Rajasthan, Karnataka, Andhra Pradesh etc.

3.1.2 Botany

Family : Leguminosae 2n = 22

Pigeon pea [*Cajanus cajan* (L) Mill sp. Synonymous : *Cajanus indicus* spreng] is an annual crop but can be grown as perennial. It has a very tap root and mass of fibrous lateral roots with almost spherical nodules. The height of the plant is 1 – 4 m. The inflorescence is raceme. Pigeon pea is self-compatible and most frequently self pollinated. But 5-40% of the crossing can occur depending upon the frequency of insect visit. The pods are flattened with pointed beak, slightly hairy, usually 4-5 cm long. Seeds are usually spherical or oval and white, brown red or specked with a white hilum. According to International Rules of Botany classification *Cajanus cajan* (L) Mill sp. has been adopted for the species known earlier as *Cajanus flavus* and *Cajanus bicolor*. These two species were formerly described under *Cajanus indicus* spreng.

Classification

Depending upon the height, habit, time of maturity, colour, size and shape of the pod and seeds the pigeon pea is classified in two categories :

(i) *Cajanus cajan* var *bicolor* : Late maturing plants, grow taller, freely branched, bearing flower with yellow colour having purple streaked at the shoot. The pods are relatively longer dark coloured with 4 or 5 seeds.

(ii) *Cajanus cajan* var. *flavus* : Short duration plants, shorter in height, bearing yellow flower at severl points along the branches. The pods are shorter having 2 to 3 seeds in them.

3.1.3 Climate

Pigeon pea needs moist and warm weather during seed germination (30-35°C), slightly lower temperature during vegetative growth (20-25°C), but about 15-18°C during flowering and pod-setting, however, at maturing it needs higher temperature of around 35-40°C. Water logging, heavy rains, frost are very harmful for the crop. Hailstorm or rain at maturity damages the entire crop. It can be grown on any type of soil, but sandy loam to clayey soil is the best. Cloudy weather and excessive rainfall at flowering time also damage red gram to a great extent.

3.1.4 Agronomic Management

3.1.4.1 Selection of land

Field selected for the seed production of pigeon pea should not have been used for the same cops in the previous year, unless the same variety was grown using a certified seed. The soil should be light, well drained and with a neutral soil reaction to obtain higher yield. It can be grown in variety of soil sandy to black cotton soil. The successful cultivation of red gram needs fertile and well-drained loamy soils. Saline soil, alkaline soil and water-logged soils are unfit for its production as these problem soils adversely affect the nodulation process. It is very deep-rooted crop, so it can be grown in fairly hard soil.

Generally a deep cultivation by ploughing, followed by harrowing and levelling may be sufficient for its production. The tillage practices do not affect significantly to its production, so conventional tillage will be sufficient. The land should be free of volunteer plant.

3.1.4.2 Sowing parameters

Sowing of pigeon pea in first week of June is recommended for higher seed yields, but in case *rabi* crop the optimum time of sowing of 2nd fortnight of September. It was observed that both types of pigeon pea (indeterminate and determinate) yield reduced with delayed sowing. Delayed sowing resulted decrease in radiation use efficiency (RUE). RUE positively correlated with pigeon pea seed yield. But on the contrary, some scientist believed that delayed sowing affected growth period of indeterminate type where as delayed sowing had no effect on determinate types. In frost prone area like Haryana and Punjab sowing late in season or with onset of monsoon cause drastic reduction in yield. If the crop sown earlier the reproductive growth occurs during period of heavy rainfall causing heavy flower and pod drop.

Sowing is done either with seed drill or hand tyne or by plough in furrows. The depth of sowing is 5 cm.

For long duration variety spacing of 120 cm x 60 cm has found to be best. But for short duration variety 60-75 cm x 25-30 cm is sufficient. For a extra early variety as well as those having compact growth habit, however, row spacing of 30-45 cm with population of 1 lakh plants/ha produce best result. It has been observed that cultivation of pigeon pea at a spacing of 25 cm between plants on ridges of 75 cm apart and application of 15 kg N + 17 kg P/ha increased the seed yield of improved cultivar 'Bahar' by 4.60 q/ha from 17.8 q/ha with local practice, but it has no effect on local variety. Actually the improved practices increased the number of seeds per pod in 'Bahar'.

About 10-15 kg of seed is required to sow an area of one hectare.

Cultivator should obtain nucleus/breeder/foundation seed from source approved by seed certifying agency. Some research findings also revealed that the influence of seed size and seed coat colour on seed quality of pigeon pea. Seeds of CO-1 varied size and colour-large, dark tan and tan-coloured seeds were superior to small and off-coloured seeds in their vigour retting, germination percentage, root growth, nodulation and seed yield. Flowering was accelerated by 8-10 days.

Germination and seedling vigour are significantly influenced by seed size but not by days to maturity. Small and medium seeds genotypes had better germination and mobilization efficiency than the bold seeded genotypes, whereas bold seeded genotypes had higher vigour index.

3.1.4.3 Seed treatment

Fungicide treatment: Seed should be treated with fungicide like Thiram + Bavistin (1 : 1) @ 2.5 g/kg of seed. After treating the seed with those fungicides, appropriate *Rhizobium* culture is to be used.

Rhizobium **inoculation**

Pigeon pea *Rhizobia* are of the cowpea cross inoculation group. The bacteria are usually slow growing with little gum production but a few fast growing strains have also been found. Soils with high do not respond well to inoculation treatment. However, light textured soil with low organic matter content, newly reclaimed eroded and flooded soil show positive response to culture. Seed inoculation with *Rhizobium* culture for legumes which is grown first time in the field is necessary. *Rhizobium* inculcation may increase seed yield and protein content of seed.

In a rainfed trial conducted during *kharif* season with a pigeon pea/long duration variety seed inoculation with *Rhizobium* strain CC-1 or seed treatments with 12% (W/W) SSP was found to increase the number and weight of pods/plant and thus increases the seed yield.

Growth regulator treatment

Pigeon pea cultivar Gwalior-3, Bahar and IPCL-1 seeds were treated with various concentrations of IBA, GA, BA, B-9 or ethephon. High concentration of ethephon (200-400 mg), B-9 and IBA (100 mg) and GA (200 mg) inhibited seed germination in some cultivars. Seed yields were generally decreased by growth regulator concentration higher than 10-20mg IBA and B-9 and 50 mg ethephon, while only 50 mg BA/lt. in Gwalior-3 and 100 mg BA/lt. in Bahar increased the yield significantly.

3.1.4.4 Isolation

In general pigeon pea is a self-pollinated crop but some partially cross-pollination is also occurred. Although the anthers burst before the flowers open, considerable cross-fertilization may take place, via bees and other insect pollinators.

Natural crossing to the extent of about 65% has been reported (Agarwal, 1995). According to Purseglove, however, cross-fertilization in pigeon pea varies from 5-40% and is normally about 20%.

The seed crops of foundation seed and certified seed classes should therefore be isolated at least by 200 m and 100 m, respectively, from other fields growing the crop.

3.1.4.5 Nutrient management

The crop is generally grown on marginal and sub-marginal lands of poor initial soil fertility. The crop is heavy feeder on the soil. As a consequence, the crop fails to produce its potential yield. Hence, the care should be taken that this seed crop does not suffer from nutrient deficiency.

Pigeon pea meets a major part of its nitrogen requirement through symbiotic nitrogen fixation with help of nodule bacteria thrives in their roots under normal condition. So, their nitrogen depends largely on the effectiveness of symbiosis which on the other hand may vary with the strain of nodule, variation in soil conditions and others. But it is well established that the potential yield can be achieved if the crop feeds on symbiotically nitrogen alone. Seed inoculation alone gave significant increase in primary as well as secondary branches. During initial stage a starter dose of 20-30 kg N/ha is recommended. To mitigate the effect of water-logging on the growth and productivity, top dressing of nitrogenous fertilizer has also been found to be effective.

To ensure the symbiotic nitrogen fixation, application of phosphorus on the basis of soil test value is must. The rate of application depends on area, season and cropping system etc. On the average 50-60 kg P_2O_5/ha has been found to be optimum. On P-deficient soil, pigeon pea received 80 kg P/ha as SSP, Udaipur rock phosphate (URP) or both in 1 : 1 gave a yield of 928, 695 and 853 kg/ha respectively, compared with 356 kg/ha without P application. Yields with 40 kg P/ha as SSP, URP and both in 1 : 1 ratio were 726.611 and 687 kg/ha, respectively.

Pigeon pea does not respond well to potassium application due to higher potassium content in Indian soil and proliferation of root system resulting absorption of potassium from deeper soil. Application of potassium @ 40 kg/

ha may be recommended for deficit soil and to higher monetary return from application of nitrogen and phosphorus.

In soil having low pH, application of calcium gives better result due to its definite effect on nitrogen fixation. Sulphur applied @ 20-25 kg/ha resulted in higher seed yield. Pigeon pea shows a high degree of susceptibility to Zn deficiency depending upon place and variety. Soil application of 2-4 ppm Zn or foliar spray of 0.5% $ZnSO_4$ with 0.05% lime have proved to be effective in controlling Zn deficiency. Pant A-3 and Prabhat are tolerant to Zn deficiency than other cultivar. Application of Sodium Molybdate and Borax are also found effective in improving seed yield of pigeon pea.

Application of organic manure @ 4-8 t/ha in sandy loam soil increases the seed yield to a greater extent. It increases the water holding capacity of light soil besides supplying plant nutrients. But for general cultivation or seed production application of FYM @ 1 t/ha is effective.

The fertilizer may be drilled at the time of sowing; placing it 10-15 cm deep is to the side of the seed. Band placement of fertilizers yields better results than the traditional broad casting method.

3.1.4.6 Roguing

Rogue the off-type plants and diseased plants affected by wilt, leaf spot and stem canker, yellow mosaic virus and sterility virus from seed field from time to time, as required. It enhances the seed purity. So, for seed crop roguing is done in three stages:

(i) Pre-flowering stage-depending on leaf and stem character.

(ii) After flowering-depending on inflorescence character

(iii) At maturity – depending on grain character.

3.1.4.7 Weed management

Pigeon pea is very sensitive to weed competition in the first 60 days of growth because of its early slow growth rate and wide spacing. The weeds commonly found in pigeon pea fields are *Commelina benghalensis*, *Phyllanthus niruri*, *Amaranthus viridis*, *Digeria ravens*, *Euphorbia hirta* (broad leaves), *Digiteria sanguinalis*, *Dactyloctanium aegyptium*, *Cyndon dactylon*, *Echinocloa colona* (grassy weeds). One to two hand weedings or hoeings when the crop is 30 days and 60 days old should done. If necessary, pre-emergence application of Alachlor @ 2 kg/ha or Pendimethalin @ 1 kg/ha has also been found effective.

3.1.4.8 Water management

Pigeon pea is a deep rooted crop and it can tolerate drought. But for getting potential yield pigeon pea may require one or two light irrigations prior to the onset of monsoon. During irregular rainfall situations and prolonged dry weather conditions, one irrigation at the time of flowering and another at the seed development stage are necessary.

3.1.5 Natural out crossing

The frequency of natural out crossing of two partially clistogamy mutant lines (ICPL 87047 and ICPL 87154) was studied at two locations in Sri Lanka and 3 locations in India. It was found that natural out crossing in the mutant lines in Srilanka ranges from 0.14 to 1.33% in comparison with 6.34 to 19.6% in controls. In the Indian environments, natural out crossing ranged from 0.16 to 2.67%. It was suggested the mutant may be of considerable economic importance in pigeon pea improvement and seed production programme.

3.1.6 Maturity

Pigeon pea seeds acquire the ability to germinate after 23-26 days after anthesis (DAA) and attain maximum potential (96-98% germination) at 45-49 DAA. It was also observed that all the genotypes attain physiological maturity at 45-49 DAA and harvestable maturity at 51-55 DAA. It was suggested that complete loss of colour (green) of pod was found useful indicator for judging the optimum physiological maturity stage in the field whereas the harvestable maturity is indicated by brown pod.

In an experiment, seed quality at physiological and harvest maturity of pigeon pea cv. AL15 and T21 was assessed by Khatra *et al.* (1997). It was noticed that the seeds attain maximum dry weights, germination potential and seedling dry weight at 35 DAA in AL 15 and 46 DAA in T 21, denoting their physiological maturities. It was observed that harvesting at physiological maturity appears to be feasible.

3.1.7 Harvesting

Pigeon pea poses non-synchronous maturity and in determinate growth. Cold weather and most soils can extent its maturity even after bearing has started. It is however, considered optimum for harvesting when about 75-80% of the pods start drying and change colour to from green to brownish yellow. The plants are cut as close as ground level.

After drying the plants are threshed either by passing the stone roller, tractor or by beating with heavy wooden sticks. After threshing the threshed materials are winnowed to get cleaned seed.

3.1.8 Yield

Seed yield largely depends on variety and management practices, average seed yield obtained from one-hectare range from 1-2 t/ha.

3.1.9 Storage

Seeds with 8-10% moisture are stored at farmers, traders and Government levels in various storage structures. Seeds may be fumigated. However, research findings revealed that there was no significant different in fumigated and non-fumigated stored samples except in case of germination. During storage a sharp decline in germination has been observed (90-68%). Fumigation by Japanese mint has also been found to be effective for germination. Moreover, there is no adverse effect on nutritive value of pigeon pea if it is stored in air tight container and fumigated with Japanese mint oil. Studies on the efficacy of ten oils viz. sunflower, castor, mustard, safflower, groundnut, palm, sesame, neem, karanja and corn each of 5, 7.5 and 10 ml/kg of pigeon pea against pulse beetle (*Callosobruchus chinensis*) were carried out with the objectives of testing their effects on the adult mortality and oviposition upto 100 days after treatment (Khatra *et al.*, 1993). It was observed that neem, castor, karanja and corn oils retain their toxicity to adult's upto 100 days of storage at 1% level.

It was also reported that percentage treatments using inert day (5 g/kg) was effective against pulse beetle (*Callosobruchus chinensis* L.) damage upto 12 months of storage due to total mortality of the released adults, without adversely affecting germination of pigeon pea.

Table 3.1: Effect of inert day on adult mortality, oviposition and adult emergence of pulse beetle, under inoculated conditions.

Treatment	Adult mortality (%)		Oviposition(%)		Adult emergence (%)	
	6 months	12 months	6 months	12 months	6 months	12 months
Pigeon (T)	100	100	-	-	-	-
Pigeon Pea (C)	-	-	100	100	96	96

Note : (T) = Treated with inert day (5g/kg seeds) (C) = Untreated control
Source : Hussaini and Ramesh Babu (1993)

A newly developed non-conventional dust, attapulgite-based day dust (ABCD) was effective as pre-storage treatment (at 0.5% w/w) for about one year in controlling pulse beetle. Pod files of pigeon pea can be controlled with a spray of thiodan @ 1.25 l/ha or nuvacron @ 625 ml/ha in 250 lt of water. Pod bugs, plume moth, caterpillars and cow bug can be controlled effective with a spray of 62.5 ml of nuvacron or 25 kg of 10% serin dust/ha. The seed crop may be dusted with 10% folidol at 25 kg/ha to control tussock caterpillar and leaf roller.

Table 3.2: Popular and released variety of pigeon pea

Variety	Source	Year of release	Yield (q/ha)	Days to maturity	Area of adoption	Characteristics
BDN-2	MAU	1978	10-12	150-160	Madhya Pradesh	Indeterminate, white seed, tolerant to wilt
UPAS 120	GBPUAT	1976	11-15	125-150	Uttar Pradesh	Early maturing
Hy 3c	IARI	1982	17.0	170-180	Andhra Pradesh	Bold and white seed
LRG-30 (Palandu)	ANGARU	1982	17-19	170-180	Andhra Pradesh	Medium maturity
Cll	PKV	1982	15-20	190-200	Maharashtra	Wilt tolerant
Vishakha 1 (TT6)	BARC	1983	10-17	130-140	Central & South Zone	Compact, medium bold seed
T15-15	GAU	1985	15-20	200-210	Gujarat	Medium maturity
CO 5	TNAU	1985	8-12	100-115	Tamil Nadu	
Manak(H 77-216)	CCS HAU	1985	18-20	120-130	NWPZ	Early maturing
Pusa 84	IARI	1985	15-16	140-150	NWPZ	Determinate type
Bahar	RAU	1986	20-25	230-250	Bihar, Uttar Pradesh	Resistant to SMD (sterility mosaic disease), seed bold and pale brown
Pragati (IPCL87)	ICRISAT	1986	11-19	116-125	Central and South Zone	Determinate type, seeds light brown
Maruti (ICP8863)	ICRISAT	1986	10-12	155-160	Andhra Pradesh	Resistant wilt
TTB 7	UAS	1988	15-17	160-170	Karnataka	Semi-spreading
Pusa 33	IARI	1988	18-20	120-150	NWPZ, Central Zone	Indeterminate type
Jagriti (ICPL151)	ICRISAT	1989	18-20	120-140	NWPZ, NHZ, CZ	Determinate type, seed cream colour
JA4	JNKV	1991	16-18	180-200	Madhya Pradesh	Tolerant to wilt and SMD
GujratTur 100	GAU	1992	16-18	120-130	Gujarat	White and bold seed
Birsha Arharl	BAU	1992	10-15	180-200	Bihar	Resistant wilt
Asha (ICPL 87119)	ICRISAT	1993	16-18	160-170	Central and South Zone	Resistant wilt and SMD, indeterminate

Variety	Source	Year of release	Yield (q/ha)	Days to maturity	Area of adoption	Characteristics
Sharad (DA 11)	RAU	1993	18-20	240-250	Bihar	White seed resistant to *Alternaria* blight and SMD
Pusa 9	IARI	1993	22-26	210-248	NEPZ	Tolerant to SMD and *Alternaria*, indeterminate, seed brown
CO 6	TNAU	1993	8-10	170-180	Tamil Nadu	Indeterminate
Sarita (ICPL 85010)	ICRISAT	1994	10-12	130-140	Andhra Pradesh	Determinate
Jawahar (KN 7)	JNKV	1996	18-20	173-180	Central and South Zone	Tolerant to wilt, seeds dark brown
Narendra Arhar 1	NDUAT	1997	20-22	240-260	Uttar Pradesh	Resistant to SMD and tolerant to wilt and PB
Amar (KA 32-1)	CSAUAT	1997	16-20	250-270	Uttar Pradesh	Resistant to SMD and tolerant to wilt, seed brown
Paras (H 82-1)	CCSHAU	1998	15-20	133-145	Haryana	Indeterminate
Azad (K 91-25)	CSAUAT	1999	20-22	250-260	Uttar Pradesh, Bihar	Resistant to SMD
Laxmi (ICPL 85063)	ICRISAT	2000	18-20	160-200	Andhra Pradesh	Suitable for pre-*rabi*
AKT 8811	Akola	2000	13-14	145-150	Maharastra	Indeterminate
Pusa 992	IARI	2002	18-20	130-140	NWPZ	Indeterminate
MA 6	BHU	2002	20-23	248-267	Uttar Pradesh	Spreading type
M13	BHU	2004	20-22	250-260	NEPZ	Tolerant to wilt

Source : Handbook of Agriculture, 2006.

Recently released varieties / hybrids of pigeon pea

Variety	Duration (days)	Yield (t/ha)	Area of adoption	Salient features
CO 9	170–180	1.7	Tamil Nadu, Odisha, Karnataka, Andhra Pradesh and Telangana	Suitable for rainfed irrigated conditions, seed weight 9.9/100 g, protein content 23.65%, moderately resistant to wilt and SMD diseases, maruca and pod fly.
Bheema	160–165	1.9	Madhya Pradesh, Maharashtra, Gujarat, Chhattisgarh	Suitable for irrigated and rainfed areas during *kharif* season, protein content 23.01%, resistant to wilt and sterility mosaic diseases.
MPV-106 (RANI)	165–174	1.55	Tamil Nadu, Karnataka, Andhra Pradesh, Telangana and Odisha	Suitable for rainfed with high fertility for *kharif* season, protein content 25.87%, moderately resistant to wilt.
Pant Arhar-6	151	1.67	Punjab, Delhi, Haryana and Western Uttar Pradesh	Suitable for rainfed and irrigated conditions of *kharif* season, protein content 25.87%, resistant to *Phytophthora* stem blight, resistant to wilt and sterility mosaic diseases, moderately resistant to pod borer (*Helicoverpa*), maruca and *Apion clavipes* and bruchid.
Telangana Kandi	150–165	1.7	Tamil Nadu, Karnataka, Andhra Pradesh, Telangana and Odisha	Suitable for rainfed/irrigated conditions during *kharif* season, moderately resistant to *Fusarium* wilt and moderately tolerant to *Helicoverpa armigera*.
IPH 15-03	153	1.6	Punjab, Delhi, Haryana and Uttar Pradesh	Suitable for rainfed/irrigated areas for normal *kharif* season, resistant to *Fusarium* wilt, SMD and moderately resistant to *Phytophthora* blight.
AL 882	132	1.47	Punjab	Suitable for irrigated conditions, tolerant to PSB and resistant to CLS and sterility mosaic.

References

Aggarwal, R. L. (1995). Seed Technology. Oxford and IBH Publishing Co., New Delhi, 2nd Edition. P. 145.

Hussaini, S.H. and Ramesh Babu (1993). Effect of inert clay, fungicides, plant oils and insect growth regulators on the development of *Callosobruchus chinensis* and viability of pulse seed. Seed research. Special vol. 1 (S.P. Sharma ed.). Indian Society of Seed Technology, New Delhi, 1993, p. 306.

Khare, D. and Satpute, R.G. (1999). Influence of days on maturity and seed size on germination and seedling vigour in pigeon pea. Seed Research 27 (2) : 170-173.

Khatra, V.M., Kachare, B.V. and Mote, U.N. (1993). Effect of vegetable oils on mortality of pulse beetle in pigeon pea seeds. Seed Research. 21 (2)) : 78-81.

Khatra, V.M., Kaur, A. and Singh, G. (1997). Comparison of seed quality at physiological maturity and harvest maturity of two pigeon pea (*Cajanus cajan* L. millsp) cultivars. Acta-Agrobotanica 50 (1-2) : 49-52.

Saxena, K.B., Jayaskra, S.J.B.A. and Ariyaratne, H.A. (1994). Frequency of natural out crossing in partially clistgamous pigeon pea lines in diverse environments. Crop Science 34 (3) : 660-662.

Singh, K. and Karkalaya, B.L. (1992). Seed germination, seedling growth and yield of pigeon pea as influenced by seed treatment with growth regulator. Annals of Agril. Research 13 (1) : 12-18.

Solaniappan, U.P. and Singh, S.N. (1986). Effect of seed treatment soil and foliar fertilization of N and P and yield attributes of pigeon pea grown under rainfed condition. Indian Journal of Agronomy 35(3):234-237.

Tripathy, H.P. and Singh, S.N. (1986). Effect of some management practices on growth and yield of pigeon pea at Tusshi, U.P., India. International Pigeon pea Newsletter 5 : 25-26.

3.2 Greengram and Blackgram

Greengram (*Vigna radiata* L., Syn. *Phaseolus aureus*) and blackgram (*Vigna mungo* L., Syn. *Phaseolus mungo*) are botanically very similar pulses. Blackgram can be distinguished from green gram by its much shorter, stout, very hairy pods and longer oblong seeds, which vary in colour from blackish to olive. Both crops are commonly grown in several tropical Asian Countries. They thrive best under a hot dry climate. Greengram is comparatively less tolerant to drought and humidity than the blackgram. Among all the pulses grown in India, blackgram is highly prized for its nutritious value and food uses.

3.2.1 Species and cultivars

About 2000 strains and botanical varieties of green gram are known, which are grouped as follows :

i) *Vigna radiata* var. *sublobata* (Roxb.) verdcourt (syn. *Phaseolus sublobatum* Roxb.); *P. trinervis* Wright & Am., *V. opisotricha*, A. Rich; *V. brachycarpa* (Kurtz.)

ii) *Vigna radiata* var. glabra (Roxb.) verdcourt (syn. *P. glaber* Roxb.); *P. mungo* var. glaber (Roxb.) Bak in Hookf; *P. calcaratus* var. *glaber* (Roxb.) Prain.

About 100 varieties of green gram are known in China and other Asiatic countries, which differ in growth habit, plant size, period of maturity, pod colour and size and colour of seeds. The main golden cultivars grown in the United States are *P. aureus*, and the green ones are *P. typical* or *radiata*. In India the black seeded types are known as grandis, and the brown seeded ones as bruneus.

Many cultivars of black gram are recognized in India, namely early and late maturing and small seeded with black or olive-green seed coat. *V. radiata* and *V. mungo* are both derived from *V. radiata* var. *sublobata* (Roxb.) verdc. (*P. sublobatus* Roxb., *P. trinervis* Wright & Am), which grows wild in India and Indonesia.

3.2.2 Agronomic Management

3.2.2.1 Isolation requirements and land preparation

A light textured and well drained field should be selected for the seed crops of green gram and black gram. The same kind of crop should not have been grown in the previous two seasons unless the previous crop was of the same variety and the seed was certified.

According to Purseglove, green gram is fully self-fertile and almost entirely self-pollinated. Therefore, isolation distance to prevent mixture at harvest about 3 m would be sufficient. Agrawal (1995), however, recommended an isolation of 20 m for foundation seed crop and 10 m for certified seed crop to avoid mechanical admixtures. Since pollen shedding takes place long before the petals open, self pollination is the rule in this pulse.

Fields should be prepared well by ploughing and one or two harrowings followed by leveling. If the seed crop is being raised after wheat or paddy (in the South), no land preparation should be necessary, and the crop can be seeded after one pre-sowing irrigation.

3.2.2.2 Planting and cultivation

Green gram and black gram are usually taken as *kharif* crops. They can, however, be sown either in the second week of February or in April soon after the wheat harvest. In the Southern parts in India, these two pulses are also grown as *rabi* (winter) crop in lighter soils with moisture retentive characteristics. They can also be grown in rice follows after paddy. Nucleus/breeder/foundation seed obtained from an approved certification agency should be used for sowing. The recommended spacing for green gram and blackgram seed crops grown in different seasons are as follows –

Kharif and *Rabi* crops	-	Row to row 30-45 cm
		Plant to plant 7-10 cm
Spring and Summer crops	-	Row to row 20-25 cm
		Plant to plant 7-10 cm

3.2.2.3 Fertilization, irrigation and weed control

In addition to 5 tons of manure, green gram and black gram crops require about 20 kg N and 35-40 kg P per hectare during *kharif* season, on medium to low fertile soils preferably. In Zn-deficient soils, a foliar spray of 0.5% $ZnSO_4$ + 0.5% lime solution has been recommended.

Spring and summer seed crops of green gram and black gram will require frequent irrigations to maintain adequate soil moisture during crop growth. *Kharif* crops may not require irrigation in areas of assured rainfall. One or two irrigations may be required if there is a prolonged dry period.

The fields of seed crops must be kept clean and free of weeds by one or two weeding or hoeing operations in the early stages of growth. A spray of 1 kg treflan, in 1000 litres of water per hectare at the time of final land preparation is recommended to check weeds in the seed plots.

3.2.3 Control of Pest and Diseases

The aphids and white flies of green gram and black gram may be controlled with a spray of 625 ml metasystox in 625 litre of water per hectare, which may be repeated at 15-20 day intervals if necessary. Dusting with 10% sevin of folidol at 25 kg/ha controls caterpillars and pod borers.

Anthracnose and Cercospora diseases can be checked with a spray of copper oxychloride at 1.25 – 2.50 kg in 625 litre of water/ha as soon as initial symptoms of the disease appear. For dry root rot, soil is drenched with copper oxychloride at 5 kg/ha in 250 lit of water. Powdery mildew can be controlled with 25 kg sulphur dist/ha or a spray or 25 kg wettable sulphur or benlate at 150 kg/ha. Rusts can be controlled with a spray of wettable sulphur at 2.5 kg/ha or 2.5 kg fine sulphur dust/ha at 15 days interval. Spraying with 0.1% lindane, 0.04% monocrotophos, or 0.07% endosulfan can effectively control stem-boring maggots, aphids, jassids and white flies of the greengram and blackgram crops.

Colletotrichum dematium has been reported to be seed borne in most legumes, and scientists have shown than *C. dematium* is also seed borne in greengram and blackgram in Southern Africa. Eight fungicides, that is benomyl, captab, fludioxonil, imazil/iprodione, manocozeb, thiram and a commercial *Bacillus subtilis* isolate were evaluated for their efficacy in reducing in vitro mycelial growth of *C. dematium*. Seeds were artificially inoculated with *C. dematium* treated with captan, benomyl, thiram and mancozeb at the recommended rates and hot water and the percentage germination and infection decreased in vivo.

3.2.4 Roguing

Off-types and insect affected and diseased plants of greengram, blackgram (blight, mosaic, anthracnose affected plants) should be rogued from time to time.

3.2.5 Harvesting and Threshing

Summer and spring seed crops of greengram and blackgram are ready to harvest when the pods turn black. The *kharif* crop is harvested when most of the pods have turned black. Harvesting practices vary in different regions. In greengram, the fruiting process extends over a long period of time. Hand picking of pods, therefore, is a common practice. In certain varieties 75% of the crop can be harvested at first picking and the remainder about 10 days later. The plants along with the roots are pulled and stacked for a week in the sun on a threshing floor. After the pods are thoroughly dried, they are threshed by beating with sticks or trampling by oxen.

Many investigators reported the effects of sowing time and stage of harvest on seed yield and quality of mungbeans. The harvesting of seed crops after 2-3 weeks of pod initiation produced higher seed yield compared to harvesting after 4-weeks. Seed yield and standard germination were significantly higher when the seed crop was sown in July as compared to August sowing, although seed weight was higher in the later.

3.2.6 Post Harvest Operations

3.2.6.1 Drying and cleaning

Small quantities of seed are normally dried by spreading on a drying floor, which if covered should be well ventilated. In humid regions artificial dryers can be used to dry seed, the air temp in this case should not exceed 35°C. Scrupulous cleaning of the grains required to avoid storage pests. Winnowing in natural wind generally performs cleaning of the threshed produce. The cleaned seed is cried to 8-10% moisture before storage. Necessary precautions must be taken to avoid mechanical admixtures during harvesting and all post harvest operation.

3.2.6.2 Seed treatment and storage

The seeds can be treated with captan or thiram to protect them against anthracnose (*Colletotrichum lindomuthianum*), wilt (*Fusarium oxysporum*) and damping-off disease (*Pythium* spp.). Streptomycin sulphate controls bacterial blight (*Xanthomonas vignicola*). According to Kelley, the best way to combat seed borne disease of pulse crop is to use only healthy seed from a clean crop. Studies conducted in India showed that, seeds of pulse treated with malathion and stored in polythene bags retain their viability over a period of 8 months. The powdered leaves of *Cymbopogon flexuous* (2-4% by weight) gave adequate protection to pulse seeds against bruchids (*Callosobruchus maculates*).

3.2.7 Some Essential Points about Seed Production

"Seeds be viable", says our holy scripts. In the dynamic system of crop evolution the seed technology and production centre has to play a pivotal role for the viability of seed. It has to cater to the needs of the farmers. After the release and notification of a variety the seed has to be multiplied in sufficient quantity. So seed production takes 3-4 generations (breeder/foundation/certified) before it reaches to the farmers. The seed technology unit concentrates of maintenance of original characteristics of the variety and ensures that the variety does not degenerate during the multiplication cycle. The seed technology research also assures good seed quantity, better seed health and the best seed quality in terms of physical and genetic purity.

3.2.7.1 Quality of seed

Availability of quality seeds of improved cultivars is considered crucial for realizing productivity and adoption of cultivars in different agro-climatic conditions. The quality of seed along is known to account for at least 10-15% increase in productivity (ICAR-1993). The good quality seed should have the following characters :

- Genetic purity, uniformity should be conformed to standards of particular cultivar
- Disease free, viable seed
- Free from admixtures of other crop seeds, weeds and inert matter
- Acceptable uniformity with respect to size, shape and colour.

3.2.7.2 Maintaining seed quality and purity at farmers level

- Avoid delayed sowing for seed production as it may produce poor quality seed.
- Remove all off-type plants before flowering. Even among the uniform looking plants, the late flowering plants should be removed to ensure that there is no drift of population towards later maturing plants.
- Prevent mechanical mixing and damage to seed. Care should be taken at the time of harvest to separate any off-type plants from the harvested bulk that might have been missed during roughing.

3.2.7.3 Seed treatment and seed production

Losses due to insect infestations and fungal damage can be heavy, unless effective control measures are maintained constantly. An effective control programme includes sanitation of seed and effective measures to keep pest and diseases out of the seed. Every certifiable seed may be treated with seed treatment chemical to keep the seed for long time without damage.

Seed production of green and blackgram in a tabular form

Particulars	Greengram	Blackgram
Land Requirements	Land to be used for seed production should be free from volunteer plants.	
Isolation	Since the pollen shedding takes place long before the potds open,	
Requirements	self pollination is the rule, therefore, an isolation of 10 meters for foundation seed, and five meters for certified seed class from fields of other varieties and same variety not conforming to varietal purity requirements for certifications if necessary.	
Cultural Practices		
Time of sowing	The crops comes up well in the *kharif* season. However, it can be sown either in the 2111 week of February, soon after the wheat harvest in April or in *kharif* season. In the South it also comes up well in the *rabi* season in light soils with moisture retentive capacity. It can also be grown in rice-fallow after paddy.	
Preparation of Land	The land should be prepared well by one to two harrowings, followed by leveling. If the crop is being taken after wheat/paddy (in South) no preparation is necessary and seeding can be done after giving one presowing irrigation	
Source of Seed	Seeds must be obtained from source approved by a seed certification agency. Seeds must be breeder/foundation/certified seed	
Seed Rate	*For kharif* & *rabi* season -15-20 kg/ha	For *kharif* & *rabi* season -15-20 kg/ha
	For spring and summer season 25-30 kg/ha	For spring and summer season 25-30 kg/ha

Particulars	Greengram	Blackgram
Fertilization	A basal application of 25 carts of farm yard manure in addition to 20 Kg N, 35-40 kg phosphorus and potash in each per ha is recommended (or *kharif* season crop.	
Method of sowing	The seed should be planted in row / in line.	
Spacing	*Kharif* and *Rabi* Crop - Spring and Summer-	3045 cm (R-R), 7-10 cm (P-P) Crop - 20-25 cm (R-R), 7-10 cm (P-P)
Irrigation	Frequent irrigation for spring and summer crop is especially necessary, the *kharif* crop does not require any irrigation. One to two irrigation maybe required if there is prolong dry period.	
Intereultural operation	It is essential to keep the weeds under suppression. Hence one/two weedings and hoeings may be done. Spraying of 1 kg Treflan (active compound) in 1000 of water per hectare on the soil, A the time of final preparation has also been found quite effective for controlling weeds.	
Insect and Disease Control	Adopt recommended IPM methods	
Roguing	The offtype plants and severely diseased plants, should be rogued out from time to time as required.	
Harvesting and	For summer and spring crop start picking when pods turn black. In	
Threshing	*kharif* crop harvesting is started when most of the pods have turned black. The threshing can be done by hand to avoid injury to seeds. After threshing and cleaning, the seeds should be dried to nine percent moisture before storage.	
Seed yield	8-10 q/ha	10-15 q/ha

Table 3.3: Popular and released varieties of blackgram

Variety	Source	Year of release or notification	Yield (q/ha)	Days of maturity	Area of adoption	Characteristics
T9	CSAUAT	1975	8-10	70-75	All zones	Wide adaptability, Erect
Pant U19	GBPUAT	1982	8-10	85	NEPZ	Resistant to YMV, erect
Pant U 30	GBPUAT	1982	8-10	75	Central zone, South zone	Resistant to YMV, erect
Jawahar Urd 2	JNKVV	1987	13	67	Madhya Pradesh	Tolerant to macrophomina, CLS (*Cercospora* leaf spot) and YMV
Jawahar Urd 3	JNKVV	1987	13	70	Madhya Pradesh	Tolerant to macrophomina, CLS (*Cercospora* leaf spot) and YMV
Pant U 35	GBPUAT	1987	10.8	75-80	Uttar Pradesh	Pods hairy
Basant Bahar (PDU1)	IIPR	1991	13	70-80	NWPZ, NEPZ, Central Zone	For spring season, ere ct, Resistant to YMV
ADT 4	TNAU	1991	8	65-70	Tamil Nadu	Tolerant to YMV
Narendra	NDUAT	1993	10	60-70	Uttar Pradesh	YMV resistant, seed medium bold and black
Urd 1 (NDU 88-8)						
Mash 338	PAU	1996	8	90	Punjab	Tolerant to YMV, seed bold
Birsa Urd 1	BAU	1996	11	80	Bihar	Tolerant to YMV
Mash 414	PAU	1996	9.6	72	Punjab	For spring season, seed blacken brown
Azad Urd 1	CSAUAT	1999	10	80	NWPZ	For spring season, YMV resistant
(KU 300) Uttra (IPU 94-1)	IIPR	1999	12	85	NEPZ& NWPZ NWPZ	For *kharif* season ,resistant to YMV
Shekher 2	CSAUAT	2001	11.3	70	NWPZ	For spring season, resistant to YMV
(KU 300) NDU 99-3	NDUAT	2003	9.5	85	NHZ	For *kharif* season, resistant to YMV
KU 96-3	CSAUAT	2003	8	73	Central Zone	For *kharif* season, resistant to YMV

Source: Handbook of Agriculture/2006/ ICAR

Recently released varieties / hybrids of blackgram

Variety	Duration (days)	Yield (t/ha)	Area of adoption	Salient features
Pant Urd 10	70-75	1.2–1.5	Jammu and Kashmir, Himachal Pradesh, Tripura and Manipur	Suitable for northern hilly regions, 100-seed weight 4.5 g, protein content 25.84%, maturity 85 days, resistant to MYMV, powdery mildew and leaf crinkle virus diseases, erect type.
Ghantasala Minumu 1	70–75	1.09 (*Kharif*); 1.33 (*Rabi*)	Andhra Pradesh	Suitable for timely and late sown conditions during all seasons, shiny black bold seeded variety, 100 seed weight 4.4 g, resistant to YMV.
VBN 9	72	1.23	Andhra Pradesh, Tamil Nadu, Karnataka and Odisha	Suitable for rice-fallow conditions of south zone, oil content 22.68%, moderately resistant to mung bean yellow mosaic virus, urdbean leaf crinkle virus, leaf curl virus and powdery mildew.
VBN 10	74	1.13	Andhra Pradesh, Tamil Nadu, Karnataka and Odisha	Suitable for *rabi* season, oil content 25.28%, resistant to mung bean, yellow mosaic virus, urdbean leaf crinkle virus and leaf curl virus.
Pant Urd 7	86	0.95	Uttarakhand	Suitable for timely sown rainfed conditions, grain protein content 23.83%, resistant to mung bean yellow mosaic virus, *Cercospora* leaf spot, powdery mildew and tolerant to whitefly and jassid.
Pant Urd 8	85	1.08	Uttarakhand	Suitable timely sown rainfed conditions, grain protein content 24.07%, resistant to mung bean yellow mosaic virus, bacterial leaf spot, *Cerospora* leaf spot, powdery mildew and tolerant to whitefly and jassid.

Variety	Duration (days)	Yield (t/ha)	Area of adoption	Salient features
Pant Urd 9	85	1.09	Uttarakhand	Suitable for timely sown rainfed conditions, grain protein content 23.90%, resistant to mung-bean yellow mosaic virus, *Cercospora* leaf spot, powdery mildew and tolerant to whitefly and jassids.
IPU 13-1	72	0.72	Different agro-climatic regions of Madhya Pradesh	Grain protein content 26.12%, resistant to yellow mosaic virus, *Cercospora* leaf spot, anthracnose, leaf crinkle, powdery mildew, bacterial leaf blight, web blight and tolerant to major insect-pests.
IPU 10-26	72	0.6	Different agro-climatic regions of Madhya Pradesh	Grain protein content 25.27%, highly resistant to yellow mosaic virus, *Cercospora* leaf spot, resistant to urdbean leaf crinkle virus, stem necrosis, web blight, bacterial leaf blight, anthracnose, powdery mildew and tolerant to major insect-pests.
IPU 11-02	71	0.7	Different agro-climatic regions of Madhya Pradesh	Grain protein content 26.42%, resistant to yellow mosaic virus, *Cercospora* leaf spot, resistant to urdbean leaf crinkle virus, anthracnose, powdery mildew and tolerant to major insect-pests.

Table 3.4: Popular variety of greengram

Variety	Source	Year of release	Yield (q/ha)	Days to maturity	Area of adaptation	Characteristics
Pusa Baisakhi	IARI	1974	8.0	60-65	NWPZ	Erect type, early, spring/ summer
PS 16	IARI	1980	8.0	60-70	NWPZ, NEPZ	Erect type, early, spring/ summer
K 851	CSAUAT	1982	14.0	75-80	All zones	Semi spreading, wide adaptability
Panna	PORS	1982	9-10	75-85	West Bengal	Early, erect type, spring/ summer/autumn
Sonali	PORS	1982	9-10	75-85	West Bengal	Early, erect type, spring/ summer/ autumn
Pant Mung 1	GBPUAT	1983	8.0	70	Whole zone	YMV resistant, seed small
Pant Mung 2	GBPUAT	1983	8-10	65-70	NEPZ, Central zone	YMV resistant, seed small
Pant Mung 3	GBPUAT	1985	11.0	75-85	NWPZ	YMV resistant, seed small
PDM11	IIPR	1987	8.5	75	Central zone	YMV resistant, for spring season
PDM 54	IIPR	1987	9.5	65	NEPZ, South zone	YMV resistant, seed medium bold
Pusa105	IARI	1987	10.0	75	NWPZ, Central Zone	YMV resistant, seed deep green
ADT 3	TNAU	1991	10.7	65-70	Tamil Nadu	YMV resistant, seed medium bold
Narendra Mung 1	NDAUT	1992	10.0	60-70	Uttar Pradesh	Tolerant to YMV, seed small
Asha	CCSIIAU	1993	12.0	75-80	Haryana	Tolerant to YMV
Pusa 9072	IARI	1995	9.0	65-75	South zone	Tolerant to powdery mildew, for rabi
Madhira 295	ANGARU	1995	14	65-70	Andhra Pradesh	Tolerant to YMV
Jawahar Mung 721	JNKV	1996	12.4	70-75	Madhya Pradesh	Tolerant to powdery mildew
TARM1	BARC/PKV	1997	8-12	85	CSZ	For rabi, seed small, resistant to PM
TARM 18	BARC/PKV	1997	12	80	Maharashtra	Seed small and shining green, resistant to PM
Pant Mung 4	GBPUAT	1997	7.5	68	NEPZ	For spring season
Co 6	TNAU	1999	10	65	Tamil Nadu	YMV resistant
Pusa 9531	IARI	2000	9	60	Central zone	For summer season
Lam 460	ANGARU	2001	12	70-75	Andhra Pradesh	Tolerant YMV
Pusa Bold (Vishal)	IARI	2000	11	62	NWPZ	Very bold seed, for summer, YMV resistant

Variety	Source	Year of release	Yield (q/ ha)	Days to maturity	Area of adaptation	Characteristics
Gangotri (Ganga 8)	RAU	2001	9.2	72	NWPZ	For *kharif*
Samrat (PDM 139)	IIPR	2001	12-15	55-60	Uttar Pradesh	For summer, moderately resistant to disease
OUM 11-5	OUAT	2002	7.3	62	South zone	For summer, moderately resistant to disease
Malaviya Jagriti (Hum 12)	BHU	2002	10-12	66	NEPZ	Early, for summer
IPM 99-125	IIPR	2004	9.8	66	NEPZ	For spring, resistant to YMV
TM 99-37	BARC	2005	11.0	65	NEPZ	For summer, resistant to YMV
COGG 912	TNAU	2005	8.0	62	South zone	For *kharif,* resistant to YMV and CLS

Source : Handbook of Agriculture, 2006

Recently released varieties / hybrids of greengram

Variety	Duration (days)	Yield (t/ha)	Area of adoption	Salient features
VGG 10-008	65–70	1.02	Tamil Nadu	Suitable for irrigated and rainfed conditions, 100-seed weight 4–4.5 g, resistant to mosaic virus and powdery mildew and leaf crinkle virus.
Pant Mung 9	75	0.96	Uttarakhand	Suitable for rainfed normal sown conditions for *rabi* season, highly resistant to MYMV, powdery mildew and tolerant to whitefly and jassid.
SML 1827	61-71	1.16	Punjab	Suitable for irrigated condition in spring/ summer seasons, resistant to YMD, CLS, BLS, anthracnose and web blight and low incidence of bean thrips.

Table 3.5: Seed treating chemical and their doses

Crop	Name of chemical	Dosages for 100 kg seed		
		Quantity of chemical (g)	Nature of treatment	Quantity of H20
Greengram	Mancozeb	70	Slurry	.500
Blackgram	Activated clay	1000	Dry	-

References

Agarwal, R.L (1995). seed Technology. Oxford and IBH pablishing Co. New Delhi and Edition.
Desai, B.B., Seeds Hand Book (Biology, Production, Processing and Storage).
Kelley & George, Encyclopaedia of Seed Production of World Crops.

3.3 Chickpea

Chickpea is the most important winter season pulse crop in India. It represents 27 percent of total area and 33 percent of total production of pulse in the country. The area and production of chickpea are 5.81 million hectare and 3.62 million tonnes in India. Chickpea is mainly used for human consumption as well as for feeding animals. Soaked grain and husk are fed to horses and cattle as concentrate and a roughage respectively. The average productivity in India is 623 kg/ha. It contains 18-20 percent protein.

3.3.1 Origin and Distribution

Chickpea has been found in Asia and now is of major importance in North India, Burma, the Middle East and the Mediterranean basin including southern Europe and Ethiopia. The crop has spread in recent time to Africa and to South America and Australia. In India the Madhya Pradesh ranks first in area as well as production of chickpea.

3.3.2 Botany

Chickpea is a small annual herb, rarely taller than 60 cm, with erect much branched stems. The plant is covered with clavate glandula hairs. Strong tap roots and lateral roots with large lobed nodules are developed. The large alternate pinnate leaves are about 5 cm long is ovate or oblong, 1-2 cm long and strongly pointed with a serrate margin. White, pink, pale blue or purple flowers are carried singly on long axillary peduncles. Five sepals are united to form a persistant calyx-tube with pointed lobes. The broad standard is clawed, the ten stamens are diadelphos, and the style is glabrous and cleared inwards. The flowers are normally cleistogamous. Pods are inflated, about 2 cm long and 1cm broad, containing one or two spherical wrinkled seeds. They are commonly brown, but some limes salmon white, red or black. Flowers are usually self pollinated, cross pollination may also occur. Flowering continue for about a month. Germination is hypogeal. Chromosome number is $2n = 16$.

3.3.3 Classification

There are two broad groups of chickpea.

1. Deshi type (brown colour chickpea – *Cicer arietinum* L.). Seeds are normally small in size. It is widely grown.
2. Kabul type (white colour chickpea – *Cicer kabulium*) Seeds are bold and attractive.

3.3.4 Climate

Chickpea is generally grown as dry crop in the *rabi* season. Sometimes it is also grown as a regularly or partially irrigated crop. It requires cool climate for its growth and development and high temperature for maturity. Severe cold and frost are injurious. Frost at the time of flowering causes flower drop. It is best suited to areas having low to moderate rainfall and a mild cold weather. Excessive rain soon after sowing or at flowering does great harm. Hailstorms at ripening cause much damage.

Table 3.6: The improved varieties of chickpea in different states are given below :

Uttar Pradesh	K-468, Radhey, Type-3, K-850, Pant-114, L-550 (Kabuli), Hima (green)
Bihar	BR-77, BR-78, ST-4, Pant G-114, BG-203, L-550 (Kabuli)
Madhya Pradesh	JG-62, JG-221, JG-74, JG-5, JG-1, Type-53, K-468, Radhey, L-550 (Kabuli)
Rajasthan	RS-10, RS-11, G-130, H-208, C-235
Haryana	C-214, H-208, S-355, Hima (green seeded), G-130, L-144, C-235
Punjab	C-235, G-130, L-550 (Kabuli)
Delhi	BG-203
Himachal Pradesh	C-235, BG-209
Gujarat	Dohad Yellow
Jammu & Kashmir	C-235, H-355, H-208
Maharashtra	Chaffa, BDN-9-3, JG-62, Annegiri-1, D-8, Warangal
Andhra Pradesh	Jyoti, Annegiri-1, JG-62, BDN-9-3
Karnataka	Annegiri-1
Tamil Nadu	CO-1, CO-2, Annegiri-1
West Bengal	B-108 (Mahamaya-1), B-115 (Mahamaya-2)
Odisha	JG-62
Assam	F-378, Radhey, JG-74, BG-203

Recently released varieties / hybrids of chickpea

Variety	Duration (days)	Yield (t/ha)	Area of adoption	Salient features
Raj Vijay Gram 204 (RVG 204)	112	1.98	Madhya Pradesh	Suitable for timely sown conditions and amenable to machine harvesting, desi type, 100 seed weight 23.1 g, plant height 59 cm, tolerant to wilt.
Raj Vijay Gram 205 (RVG 205)	114	1.73	Madhya Pradesh	Suitable for timely sown irrigated conditions, green seeded chickpea variety, 100 seed weight 20.3 g, tolerant to wilt.
Raj Vijay Kabuli Gram 111 (RVG 111)	117	1.67	Madhya Pradesh	Suitable for timely sown irrigated conditions, Kabuli type, 100 seed weight 26.12 g, tolerant to wilt.
Raj Vijay Kabuli Gram 151 (RVG 151)	113	1.77	Madhya Pradesh	Suitable for timely sown irrigated conditions, extra-large seeded Kabuli chickpea variety, 100 seed weight 54.4 g, tolerant to wilt.
GNG-2299 (Purva)	116	1.45	Assam, West Bengal, Jharkhand, Bihar, Eastern UP, Manipur	Suitable for late sown irrigated conditions, 100 seed weight 15.9 g, tolerant to wilt (*Fusarium oxysporum*).
IPC 2006-77	112	2.07	Chhattisgarh, Madhya Pradesh, Gujarat, Southern parts of Rajasthan, Bundelkhand tracts of Uttar Pradesh	Suitable for late sown conditions under rice fallow, medium small seeds (16.5 g/100 seeds), moderately resistant to wilt, dry root rot and stunt.
PDKV Kanchan (AKG-1109)	109	1.94	Maharashtra	Suitable for early sown conditions, medium bold seed, 100 seed weight 19.7 g, resistant to *Fusarium* wilt.
Pusa Chickpea 10216 (BGM 10216)	106	1.48	Bundelkhand region of Uttar Pradesh, Madhya Pradesh, Maharashtra, Gujarat	Suitable for timely sown rainfed conditions, first MABB derived variety of chickpea in the country, 100 seed weight 22.2 g, protein content 22.6%, drought tolerant.
Pusa Parvati (BG 3062)	108-118	2.15	Maharashtra, Madhya Pradesh, Bundelkhand region of Uttar Pradesh and Southern Rajasthan	Suitable for timely sown irrigated conditions during *rabi* season, grain protein content is 20.73%, highly resistant to wilt, dry root rot and stunt.

Variety	Duration (days)	Yield (t/ha)	Area of adoption	Salient features
Haryana Chana No 7 (HC 7)	127	2.41	Punjab, Haryana, Western Uttar Pradesh, Delhi, North Rajasthan, Jammu and Kashmir, plains of Himachal Pradesh and Uttarakhand	Suitable for late sown irrigated conditions during *rabi* season, grain protein content 17.23%, tolerant to *Helicoverpa armigera*.
Super Annigeri-1	104	1.86	Andhra Pradesh, Karnataka, Maharashtra and Gujarat	Suitable for timely sown rainfed/irrigated conditions, MABB derived variety of chickpea, 100 seed weight 18.4 g, protein content 22.4%, drought tolerant.
Phule Vikram	110–115	2.27	Maharashtra, Gujarat, Madhya Pradesh, Rajasthan and Uttar Pradesh	Suitable for rainfed conditions, resistant to *Fusarium* wilt.
Pant Gram 6	145	2.17	Uttarakhand	Suitable for irrigated/rainfed late sown conditions, grain protein content 22.01%, tolerant to wilt, botrytis grey mould and pod borer.
Him Palam Chana 1	151	1.62	Himachal Pradesh	Suitable for rainfed conditions during *rabi* season, resistant to *Ascochyta* blight and tolerant to wilt, root rot, stem rot and rust.
IPC 2004-98 (Desi)	141	1.43	Uttar Pradesh	Suitable for normal sown conditions, moderately resistant to wilt.
IPC 2005-62 (Desi)	118	1.02	Uttar Pradesh	Suitable for late sown conditions, grain protein content 27.25%, moderately resistant to wilt.
Sabour Chana-1	130–135	2.2–2.4	Bihar	Suitable for normal sowing under rainfed/ irrigated conditions, grain protein content 21.66%, moderately resistant to pod borer.

3.3.5 Land requirements

Select lands on which same kind of crop was not sown in the previous season unless the crop of the previous season was of the same certified variety. Land should be high and well drained. Medium type of soils are most suited for its cultivation. Soil should be free from excessive salts with near neutral in reaction.

3.3.6 Agronomic Management

3.3.6.1 Isolation requirements and land preparation

Chickpea flowers are generally self-pollinated before they open, hence an isolation of three metres from chickpea fields is important for pure seed production of chickpea.

Fields should not be prepared to fine tilth. The soil is not compacted but is left somewhat cloddy and be allowed some time for aeration.

3.3.6.2 Sowing parameters

Chickpea sowing should be done in 2nd fortnight of October. Delay sowing should be avoided for seed production. Under rainfed condition sowing should be done earlier in the first fortnight of October (Panwar *et al.*, 1980). In case, if adequate moisture is available in the seedling zone, sowing in the second fortnight may ensure better crop. In early sowings, high temperature and humidity often encourages excessive vegetative growth and incidence of wilt disease. However, irrigated farming has sufficient scope for variation in date of sowing for good seed production.

Sowing is done either with seed drill or hand tyne or behind the plough in furrows. Depth of seedling should be 3 to 4 cm. Spacing of 50-60 cm x 10 cm has found to be best for seed production. About 45 to 60 kg seed per hectare is required for seed production of chickpea depending upon the size of the seed. Obtain nucleus/breeder's/foundation seed from a source approved by seed certification agency.

3.3.6.3 Fertilization, irrigation and weed control

Application of 5 t Farm yard manure/compost along with 20 kg N and 50 kg P per hectare is required for chickpea seed production in medium to low fertile soil. Responds of 50-75 kg P_2O_5/ha have also been recorded on the soils low in phosphorus status (Panwar, 1979). We also reported that on cultivators field, application of 20 kg K_2O/ha along with 20 kg N + 40 kg P/ha increased the yield of 90-200 kg grain/ha. Foliar application of 0.2 to 0.3 per cent diammonium phosphate at 30 days after sowing and at flowering stage has been found very effective for getting better seed yield.

One presowing irrigation should be given if moisture in soil is not sufficient for germination of seed. One irrigation at 45 days after sowing and another irrigation at early pod filling stage should be given for better seed yield. Saxena (1977) indicated that one irrigation at branching (45-60 DAS) and another irrigation at pod initiation stage gave better seed yield.

One to two weedings in early stages of crop growth are necessary to keep seed field clean of weed infestation. Yadav (1983) observed that Tribunil 1.50 kg ai/ha proved to be the most effective herbicide to control weeds in chickpea plot.

3.3.6.4 Plant protection

Gram cut worm (*Agrotis ipsilon* Hufn) cuts either while plant or its twigs. Endosulfan (Thiodan) at the rate of 0.07% and Monocrotophos at the rate of 0.01% can effectively control the incidence of different insects (cutworm, pod borer etc.) of chickpea.

The wilt disease of chickpea can be control by a little late planting, use of resistants varieties (G-24 and C-235). Loosening and subsoiling during season is a good agronomic practice to increase conservation of moisture as well as reducing the incidence of the disease. *Sclerotinia* blight is caused by *Sclerotinia sclerotiorum*. It is soil as well as seed borne disease. Use of resistant varieties and crop rotation with cereals may be used as preventive measure.

3.3.7 Roguing

All the off-type plants distinguishably on the basis of plant characteristics must be removed before flowering to ensure pure seed production of chickpea. The diseased plants affected by blight and wilt should be removed from the seed field from time to time as needed.

3.3.8 Harvesting and Threshing

The chickpea crop for seed should be harvested when the seeds are fully mature. At mature stage the leaves are reddish brown and start shedding. The plants may be pulled by hand or cut with sickles and stacked in small heaps in the field to dry in the sun for a few days (1 to 2 weeks). Then there are shifted to threshing floors. Threshing may be done either by beating the plants by sticks or with the help of suitable thresher with the caution that excessive beating should be avoided so that the seed is not injured. The seed is winnowed and cleaned after threshing and dried to 8 to 10 per cent moisture content. Care should be taken to avoid mechanical admixture during these operations.

3.3.9 Seed Yield

With good management practices seed yield of chickpea varies from 15 to 20 q/ha.

References

Panwar, K.S. Faroda, A.S., Malik, D.s., Ahlwat, I.P.S., Dhingra, K.K. Rao, M.R. Singh, Dahiya, Atar and Roy Sharma, R.P. (1980). Silver Jubilee Indian Soc. Agronomy (1955-80).

Panwar, K.S. (1979). All India Rabi Pulse Workshop 19-23 Sept. HAU., Hissar.

Saxena, M.C. and Singh, H.P. (1977). Expt Tech Bull 101:1-77, G.B. Pant Univ. Agric. and Tech. Pantnagar, India

Yadav, D. S. (1983) Report of the Research work on rabi pulses, NDUAT, Faizabad, U.P.

3.4 Lentil

Lentil (*Lens culinaris* L.) is an important oldest nutritious pulse crop of winter season. Lentil is mostly used as 'dal' and different delicious dishes. The splited seeds are also used in making soup and porridge. Green pods are used as vegetables. The leaves and stalks are used as cattle fed. Being a leguminous crop, it restores the soil fertility. It is grown in almost all parts of the country as an intercrop or pure crop. It is an important crop of dryland agriculture. It is mainly grown on low lying, inundated and marginal lands. Lentil is also grown as 'utera' crop in low-lying rice growing areas. The general productivity of lentil is very low as it is grown under marginal conditions. High yields can be obtained with use of improved high yielding varieties and agronomic practices. Lentil is more cold tolerant than chickpea and pea.

3.4.1 Origin and distribution

Lentil is believed to be originated from South Western Asia and Mediterranean region. The valleys in the Hindukush mountains in India are also considered as its native place. It then spread to Europe, India, China, Ethiopia, Africa, USA, Mexico, Peru, Argentina and Columbia. The important lentil growing countries of the world are Italy, Greece, Egypt, India, Tajakistan, Canada, Bangladesh and Nepal. In India it is cultivated throughout North and Maharashtra. The leading states include Madhya Pradesh, Uttar Pradesh, Bihar, Maharashtra, Assam, Odisha, Haryana and West Bengal.

3.4.2 Botany

Lenti (*Lens esculenta* Moench or *Lens culinaris* Medik) is a small, light green annual herb with much branched, and square stems. It has a slender tap root and a mass of fibrous lateral roots with very small, round or elongated nodules. The alternate, pinnate leaves have upto seven pairs of leaflets and a terminal bristle or tendril; they are subtended by upward pointing, linear stipules, but there are no stiples. Each sessile leaflet is oval or lanceolate and about 12 mm long; the pairs of the leaflet are not always opposite. The small white flower is tinged with purple and occurs in groups of two or three at the end of long, very slender peduncles. The calyx tube divides near its base into five narrow, pointed lobes which curve over and beyond the corolla. The standard petal is white with blue makings, while the wings and keel are white. The stamens are diadelphous and the ovary, containing one to two ovules, is surmounted by a short, curved style which is hairy along its inner surface. The pod is flattened and broad, nearly 12 mm long, tipped by a minute curved beak and with the persistent calyx around its base. The seeds are lens shaped, light red, brown or grey and often speckled with black. The hilum is minute. Germination is hypogeal. There are three types of root systems are recognized namely (i) deep roots, (ii) well branched shallow roots and (iii) intermediate types.

Classification

Lentil is under the family leguminosea with 2n chrosome number 14. Two species of genus *Lens* been identified as cultivated species, i.e. *Lens esculenta* Moench and *Lens culinaris* Medik. Species *esculenta* has been further divided into two as follows :

(i) *Lens esculenta macrosperm*a Moench, commonly known as '*masur*'. Seeds are fold and flattened about 6-9 mm diameter. Pods are large and flat. Flowers are mostly white and large.

(ii) *Lens esculenta microsperma*, commonly known as 'musri'. Seeds are small about 3-6 mm in diameter and convex. Pods are small convex. Flowers are small with violet blue to white colouration.

Growth stages

Legumes, like many other crops are sensitive, to variation in length of the day, and day and night temperature. The factors of climate influence different growth stages of the crop and thus affect the yield very much. Growth and development involve important changes in morphology and reproductive state. There are five growth stages have been identified in lentil crop i.e. seedling stage (plant is small and tender with one leaf), branching stage (plant with three fully developed leaves, 4th leaf in developing stage and primary branches begins to develop), flowering stage (plant with 10-11 leaves, 1-2 flower nodes and 11-15 vegetative nodes), pod formation stage (pod begins to develop) and maturity stage (90-95% leaves starts withering).

3.4.3 Climate

Lentil is a winter season crop. Crop needs cool temperature during growth stages and warm temperature at the time of maturity. The optimum temperature for lentil growth is 18-30°C. The plant can tolerate frost, but sensitive to high temperature. Unlike chickpea winter rains at any stage of lentil do not have harmful effect. It can be grown to the cold areas of Ladakh in Kashmir at an altitude over 3500 meters.

3.4.4 Land requirement

Select a field in which lentil crop was not grown in the previous season, unless the variety grown is the same and approved by the certified agency for varietal purity. The field should be well drained and properly leveled so that temporary water stagnation should not take place. Lentil is grown on a variety of soils ranging from sandy loam to black cotton soils. The crop is generally grown after harvesting paddy in low-lying areas. For good yield crop should be grown in well drained, and neutral loam soils. Field should be prepared soon after withdrawal of monsoon. One to two ploughing and two to three harrowings, followed by leveling are sufficient to prepare a field to the desired tilth. The soil should be weed free so that seeding could be done at uniform depth. In case of less moisture in the field it should be irrigated before final land preparation for proper germination of seed. To protect the crop from soil borne insects specially termite, apply carbofuran @ 25 kg/ha at least one week before sowing just before last ploughing.

3.4.5 Variety

Use of improved varieties of lentil seed according to suitability enhances crop yield. The following varieties of lentil have been recommended for cultivation in different states.

Uttar Pradesh : Type-36, Pant L-406, L9-12, Lens 4076, Pant L 639

Bihar : BR-25, Number 26, Pant-L406, PL 369

Madhya Pradesh : JLS-1, T36, B 77, PL 406, L9-12, Lens 4076

Punjab : L-9-12, Pant L-406, Lens 4076, PL 639

Himachal Pradesh : L9-12, Vipasha

Jammu and Kashmir : T-36, L 9-12

West Bengal : Asha (B77), B 177, Ranjan (B256), B 235, Moitreyee

North-Eastern Hills : T36, Pant L-406, B-77 (Asha), BR 25

Assam : L9-12, Asha (B 77)

Delhi : Lens 4076, Pusa 4, Pusa 6, PL 639

Gujarat : Lens 4076, PL 639

Maharashtra : Lens 4076, L9-12, PL 639

Odisha : B 77 (Asha)

Rajasthan : L-12

The important characteristics of some promising varieties of Lentil are given below :

Type 36 : Developed by local selection from Badauon in 1956; matures in 130-140 days; brown small seeded (1.5-2.0 g/100 seeds); good for late sowing; average yield 15-20 q/ha.

Pant L-406 : Developed by selection from P-495 in 1979; seed weight (2.0-2.5 g/100 seeds); suitable for the plain lands or north-east and north-west India.

BR-25: Developed by local selection; matures in 125-130 days; plants erect; grains reddish-brown; small seeded (2.0-2.5 g/100 seeds); suitable for south Bihar and Chotanagpur areas; average yield 15-20 q/ha.

Number 26 : Developed by local selection; matures in 125-130 days; average yield 20-25 q/ha; suitable for north Bihar.

JLS-1 : Developed by selection from local collection; matures in 120-130 days; grains brownish-grey (2.5 g/100 seeds); suitable for central Madhya Pradesh.

L-9-12 : Developed by selection in 1962; plants erect; grains brownish with black spots in pericarp; seed weight (2 g/100 seeds); average yield 15-20 q/ha.

Lens 4076 : Its pedigree is Pant L 234 x Pant L 639 and was identified in 1984 for North-west plains and central zones. Seeds are bold (35 g/1000 seed). It is moderately resistant to rust and is suitable for western part of UP. Average yield is 10-16 q/ha.

Asha (B77) : Medium sized plants, spreading with many branches; small seeded, brownish-grey in colour; matures in 125-130 days; protein 27%; average yield 14-16 q/ha.

B-177 : Developed by selection in 1961 from collection from Jorhat area; seeds medium sized; brownish-grey seeds; matures in 125-130 days; protein 29%; average yield 15-20 q/ha.

Ranjan (B-256) : Developed by mutation in 1976; matures in 125-130 days; semi-spreading plants; seed weight (1.8 g/100 seeds); protein 28%; average yield 15-17 q/ha.

B-235 : Branched but compact plants; small seeded; matures in 116-120 days; protein 28%; average yield 16-18 q/ha.

3.4.6 Agronomic management

3.4.6.1 Sowing parameters

Timely sowing is the key factor for realizing full yield potential of lentil. Previous findings indicated second fortnight of October is the optimum time for planting lentil (Panwar, 1978). In certain years when high temperatures persist for long duration, good yield can be obtained even if sowing is done upto mid-November. If high temperature prevails early in the season, the stand becomes thin due to *Fusarium* wilt. In Bihar, variety Pant L-406 gave maximum yield (19 q/ha) when sown in the fourth week of October at 15 cm row spacing. In West Bengal, sowing late in October or early November after the rains gave the best yields (Sen, 1982). Varietal interaction with time of sowing was also observed at Pantnagar and Bihar. Pant L-406 was found to be suitable for early as well as late sowing.

The lentil crop should be sown in lines by seed-cum-fertilizer drill or can be sown behind the desi plough. The crop is sown at row to row space of 30 cm, and plant to plant space is kept 10-12 cm and adjusted at the time of thinning. Depth of seeding should be 3 to 4 cm. Experiments at Berhampur (West Bengal) indicated that high lentil yields are obtained under high density either by establishing seeds by broadcasting or 15 cm x 15 cm plant spacing. Seed sowing at the rate of 40 kg/ha gave high yields. Under late sown conditions in central and western parts of UP, narrow row spacing (15 or 22.5 cm) has given high yields but at Pantnagar where the soil conditions are more favourable, 30 cm wide row spacing gives equally high yields. The optimum seed rate for timely sown crop for small seeded varieties is 25-30 kg/ha and for large seeded varieties is 35-40 kg/ha. Rathore (2002) indicated that optimum seed rate for timely sown lentil crop is 30-35 kg/ha. Under late sown conditions, closer spacing of 20-25 cm with seed rate of 50 kg/ha is recommended.

For seed production of lentil, nucleus / breeder / foundation seeds obtained from an approved seed certification agency should be used for sowing. Seed inoculation with *Rhizobium* culture for legumes which is grown first time in the field is necessary. *Rhizobium* should not be considered seed bacteria, they are soil bacteria. Some workers have reported that seed coat may be hostile to bacteria in some cases. Generally solid base inocula such as peat, lignite or compost provide more protection to the bacteria than to liquid lyophilized inocula (Burton, 1979b). *Rhizobia* applied to seed in a slurry with 10% sugar live longer than those applied in water.

After treated the seed with fungicide (Carbendazim 2 g/kg seed) appropriate *Rhizobium* culture is to be used. The nitrogen requirements at later stages are full filled by N fixation by inoculating seeds with *Rhizobium leguminosarum*. In USA, the inoculum at the rate of 4.4 g/kg seed is used regardless of seed size. Soil heavily infested with native highly infective non-N_2-fixing strains require greater numbers of effective inoculum rhizobia than virgin soils. Burton (1979c) reported that 10^5 to 10^6 viable *Rhizobia* per seed are required under these conditions.

3.4.6.2 Nutrient management

Addition of 5 t FYM/compost along with 20 kg N 50 kg P_2O_5 and 20 kg K_2O/ha as basal increased the growth and seed yield of lentil. Mondal *et al.* (2004) reported the highest mean seed yields (925, 959 and 918 kg/ha) was obtained when lentil was supplied with 20 kg K_2O/ha, 40 kg S/ha and *Rhizobium* inoculation, separately.

Ali (1989b) while reviewing the results of different co-ordinating centers found that significant response to 18 kg N, 46 kg P_2O_5 and 20 kg K_2O/ha in terms of 1.25, 1.10 and 1.04 q/ha seed yield respectively over control. On zinc-deficient soils, application of 25 kg $ZnSO_4$/ha gave good response. Zinc deficiency in growing crops has commonly been alleviated by foliar application of zinc sulphate (0.5% zinc sulphate with 0.25% lime). Mondal *et al.* (2004) observed that application of 20 kg K_2O/ha, 40 kg S/ha and *Rhizobium* inoculation increased the seed yield of lentil.

3.4.6.3 Irrigation management

Lentil is generally grown under rainfed condition. Lentil seldom responds to supplementary irrigation. Being deep-rooted crop, it makes fairly good use of moisture stored at lower depth of the soil. In case of limited moisture storage or ill distributed winter precipitation, crop performance could be improved by supplementing irrigation at least for two times, once at pre-flowering stage and second at pod filling stage.

At Kanpur (UP), one irrigation at early pod filling stage was most effective for lentil (Panwar and Paliwal, 1975) whereas Ojha *et al.* (1977) reported better response from one irrigation when applied at pre-flowering or post-flowering stage.

At Hissar (Haryana), two irrigations- one at 45 days after sowing and the other at early pod filling stage gave maximum seed yield and profit (Singh *et al.,* 1979). The water requirement of lentil crop is 200 mm and depth of each irrigation is 6-8 cm.

3.4.6.4 Weed management

Lentil is a slow growing crop particularly at early stages and it takes about 45-60 days to cover the land completely. Hence, the crop is infested badly by weeds. The first 30-60 days are considered critical for crop-weed competition. The losses of 25-30% are recorded in the crop, when infested with weeds. The major weed flora associated with the crop is *Chenopodium album, Fumaria parviflora, Melilotus alba, Anagallis arvensis, Asphodelus tenuifolius,* and *Vicia sativa* as broad-leaf weeds and *Cynodon dactylon* as grassy weeds and *Cyperus rotundus* as sedge. Weeds must be controlled at right stage. Two hand weeding at 30 and 60 days after sowing prove very effective. Herbicidal weed control consists of fluchloralin 0.75 kg/ha in 800-1000 lit of water as pre-plant and should be incorporated into soil (Rathore, 2002). Excellent weed control and increase in grain yield of lentil have been reported by the use of prometryne at 0.5 kg a.i./ha (Ahlawat *et al.,* 1979).

3.4.6.5 Isolation requirements

The lentil crop is generally self-pollinated. The dehiscence of anther takes place in the bud itself sometimes before the opening of flower in the morning. Hence, an isolation of three meters from other lentil fields is sufficient for pure seed production of lentil crop.

3.4.6.6 Plant protection measures

There are no serious pests that affect the lentil crop. The hairy caterpillar feed the leaves and flowers as well as pod borer make hole into the pod and grain. To control hairy caterpillar and pod borers, spray 0.04%t monocrotophos solution (1000 litres per hectare) at the appropriate time.

Wilt disease is caused by fungus *Fusarium oxyporum* sp. *lentis*. The diseased plant appears yellowish and in severe cases plant dries up and ultimately die. It affects the plant at seedling stage and at full growth stage. Seed treatment with 0.3% Benlate, use of crop rotations and resistant varieties (Pant L 406, PL 638) are helpful in controlling the disease. Do not grow lentil in field infested with wilt at least for 3-4 years.

Rust is caused by fungus *Uromyces fabae*. The affected plant shows pink to brownish pustules on stem and leaves. In advance stage, black pustules also develops. If disease is not controlled, the affected plants may wither and die. To check the disease, seed treatments, use resistant varieties (Pant L 406 and L 9-12) and spray fungicide at 15 days interval depending upon severity of disease.

3.4.6.7 Roguing

In lentil field all the off-type plants distinguishably on the basis of plant characteristics must be removed before flowering to ensure pure seed production of lentil. The diseased plants affected by blight and wilt should be removed from the seed field from time to time as needed. Weed plants particularly *Lathyrus spp.* and *Vicia spp.* Should be rogued-out from the seed field from time to time as required.

3.4.6.8 Harvesting and threshing

The lentil crop should be harvested, when plants become yellow and pods turns brownish. At this stage the lower pods are usually brown. Over ripening of pods leads to shattering, hence plant should not be over-dried. The plants may be pulled by hand, or cut and the crop is dried in sun on the threshing floor and threshed by beating the plants with sticks or by bullocks. The seed is cleaned by winnowing and then the seed should be dried to 8 to 10% moisture before storage. During threshing operation care should be taken to avoid mechanical admixtures, contact with moist soil and injury to seeds.

3.4.6.9 Yield

Generally the seed yield of lentil is very poor as it is mostly grown under abnormal situations. However, under normal conditions, lentil can produce 20-25 q/ha of seeds. Even crop cultivation by using improved technologies may yield 20-25 q/ha (Rathore, 2002).

3.4.6.10 Storage

Seeds for storage should contain less than 12% moisture. Through sun-drying the moisture content in seed can be reduced to 9%. Any seed at the stage of physiological maturity has maximum germination and vigour. After this stage the germination and vigour decline, and the rate of decline depends upon the weather conditions prevailing during ripening of the seed and post-harvest storage condition. The seed moisture at the stage of physiological maturity is generally high which is not safe for storage. The amount of moisture in the seed is probably the most important factor influencing the length of time the seeds remain viable. The storage conditions should be checked at regular intervals for cleanliness, presence of insects and moisture content of the seeds.

References

Ahlawat, I.P.S., Saraf, C.S. and Singh, A. (1979). Indian J. of Agron. 24(2) : 237-239.

Ali, Masood (1989b). Consolidated Report of Pulses. Agronomy, DPR, Kanpur (U.P.).

Burton, J.C. (1979b). In: Symbiotic nitrogen fixation in Plants. (Ed. P. S. Nutman) Cambridge Univ. Press. England, pp. 175-189.

Burton, J.C. (1979c). In: Proceedings of World Soybean Research Conference. II.Ed. F.T. Corbin. p. 89.

Mondal, S.S., Ghosh, A., Acharya, A. and Sarkar, A. (2004). Production Potential and Economics of Different rainfed rice (*Oryza*)- based paira cropping system and its effect on fertility buildup of soil. Indian J. Agronomy, 49: 6-9.

Ojha, S.N., Jha, J.N., Roy, R. K. (1977). Indian Journal of Agronomy 22(1): 54-56.

Panwar, K.S. (1978). All India Subject Matter seminar on Pulse Production Technology, 28-30 August CSAUAT, Kanpur.

Panwar, K.S. (1978). Indian Journal of Agricultural Research 11(1): 51-52.

Panwar, K.S. and Paliwal, H.S. (1975). Indian Journal of Agronomy 20(2): 21-24

Rathore, R. (2002). Techniques and management of field crop production. Published by Agrobios (India): 308-316.

Sen, S.N. (1982). Pulses and Oilseeds in West Bengal. pp 75-80.

Singh. A., Ahlawat, I.P.S. and Saraf, C.S. (1979). Progress Report of Division of Agronomy, IARI, New Delhi.

3.5 Pea

Pea is a winter season pulse crop. It is most important high yielding pulse crop. Unripe pods are used as green vegetable. It is used as 'dal'. Green seeds are canned for the use in the off-season. Pea is also grown for green fodder and it is a palatable feed for the animals. The broken grains are also used as cattle feed. It is very much nutritive and contains 22-24% protein. Pea can be used in various ways for human consumption.

3.5.1 Origin and distribution

Vavilov (1926) is of opinion that the Asiatic Centre is the primary centre of origin of pea. In Europe peas have been grown since Bronze Age. Peas were apparently cultivated in Greece in Neolithic times. In Switzerland they have been found in deposits of both the stone and Bronze Ages. It probably indigenous of the region consists of Italy and South-Western Asia eastwards of Himalayas including North India. The major parts of the area under this crop are in the Indo-Gangetic plains, Uttar Pradesh. It is also grown in other states, Punjab, Haryana, Rajasthan and Jammu & Kashmir.

3.5.2 Botany

The plant is a short-leaved, herbaceous annual, glaucous and climbing by means of leaflet tendrils. Dwarf, semi-dwarf and tall cultivars occur, depending upon number of nodes and length of internodes. The root system, apart from the tap root, is not strongly developed. The stem is weak, slender and circular in section, and hence taller varieties require support. Leaves pinnate with upto three pairs of leaf like leaflets and terminal branched tendril, stipules large and leaf like, ovate, lower portion denticiculate upto 6.5 cm long, larger than leaflets; stiples absent; leaflets ovate or elliptic upto six in number, entire or with undulate margin upto 5.5 x 3 cm. Flowers axillary solitary or in racemes of upto three flowers; bracts very small; calyx oblique, lobes unequal about 1 cm in length. Corolla white, pink or purple; keel short, incurved, obtuse; stamens diadelphous, filaments broad, anthers uniform, style falcate, flattened, bearded on inner surface, stigma minute, terminal. Pods are swollen or compressed, straight or curved on short stalks upto 15 x 20 cm with as many as 10 seeds dehiscent by two valves. Seeds angular or globose, smooth or wrinkled, no endosperm, green, grey or brown, sometimes mottled. Average seed weight is upto 250 mg. Germination is hypogeal.

The flowers, which undergo early bursting of anthers, are predominantly self-fertilized. However, natural crossing has also been observed. Chromosome Number is 2n=14.

The cultivated peas (*Pisum sativum* Linn.) are classified into two.

(i) *Pisum sativum* var. *hortense* (Garden Pea). It is also called table pea. Green seeds of this type are generally used for vegetable purpose.

(ii) *Pisum sativum* var. *arvense* (Field pea). The matured seeds of this type are used as 'dal'. This type is also used for green fodder and green manure.

3.5.3 Climate

Pea is grown as cold weather crop. It favours the cool climate and bright sunny weather which favours its growth and development. Both higher temperature and frost are detrimental for growth and development of pea plants. Pea can be successfully cultivated in temperate and semi-arid climate. High humidity during the growing season also induces the disease infestation. The optimum temperature for its growth and development is 13-18°C (Rathore, 2002).

3.5.4 Land requirements

Field selected for the seed production of pea should not have been used for the same kind of crop in the previous season unless the crop in the previous season was of the same variety of certified seeds. Pea can be grown on a variety of soils ranging from sandy loam to clay loam. In addition, the land should be well drained and neutral in reaction. It cannot be grown successfully on the soils on acidic and alkaline soil.

3.5.5 Isolation requirements

In general pea is a self-pollinated crop with a little natural crossing. An isolation distance of 20m for foundation seed class and 10m for certified seed class from other pea fields is necessary for certified seed production.

3.5.6 Variety

The cultivation of recommended pea varieties in different states are given below :

Uttar Pradesh	: Tye 163, Rachna, Hans
Bihar	: BR-2, BR 12, Swarna Rekha, Rachna, T 163, Hans
West Bengal	: B-22, BR 12, Rachna
Delhi	: Hans (L 116), Harbhajan (EC 33866), Rachna, Bonnvielle
Maharashtra	: Khaparkera, Rachna, Pant P5
Himachal Pradesh	: Kinnori, Rachna, Hans
Punjab	: PG 3, Rachna, Hans, Pant P5
Odisha	: T163, Rachna
North E.H. Region	: T 163, Hans, Rachna
Madhya Pradesh	: Rachna, T 163, Hans
Jammu & Kashmir	: Rachna, T163, Hans, Kinnori
Haryana	: Rachna, Pant P 5
Assam	: T 163, Hans, Rachna
Tamil Nadu	: T 163

The important characteristics of the promising pea varieties are given below:

Type-163 : Developed by selection from a local variety of Bulandsahar; matures in 130 days; plants tall (150-175 cm); white flowered; medium sized pods (5-6.5 cm); 4-5 seeds per pod; round white-seeded; weight (20.9 g/100 seeds); protein 21.6%, average yield 15-20 q/ha.

BR-2 : Developed by selection from local collection; matures in 145-150 days; white-seeded; 5-6 seeds per pod; round bold seeded; weight (23.5 g/100 seeds); cream coloured; average yield 16-18 q/ha.

BR-12 : Developed by selection from local collection; matures in 160-165 days; tall plants; white flowered; pods long 6-7 seeds/pod; bold seeded; cream coloured seeds; suitable for green pods; green pod yields 60-70 q/ha; average grain yield 18-20 q/ha.

Swarna Rekha (1436) : Developed by selection; matures in 120 days semi-spreading plants; white flowered; bold seeds; weight (21.6 g/100 seeds); cream coloured; round rows; suitable for green pods and to be used for pulse (*dal*); average yield 80-100 q/ha (green pod) and 15-20 q/ha.

B-22 : Developed by selection in 1963-64; matures in 135 days; small seeded; average yield 16 q/ha.

Bonnvielle : Introduced from USA; plants 100-120 cm tall; tolerant to white mildew; seed-coat shriveled; average yield 10-12 q/ha.

Hans (L-116) : Developed by selection; seed-coat smooth; round; white; small; mature 15-20 days earlier than T-136; average yield 18-20 q/ha; also suitable for late sown condition.

Harbhajan (EC-33866) : An introduced variety; very early maturing; good for green pods; small plants; pods long; weeds small; seed-coat slightly shriveled.

Khaparkhera : It is suitable for central India where the winter is short. The plants are tall, pods are short and the seed-coats are usually shriveled.

Kinnori : Developed by selection from local varieties of Kinnore, seeds are round, white; suits well in the mid and high-hills; average yield 15-25 q/ha.

PG-3 : Developed by hybridization (T-163 x Bonnvielle) dwarf; white flowered; seed-coats partly shriveled, white hard, creamy coloured and late in maturity.

Pant P 5 : Its pedigree is T 10 x T 1163 and was identified in 1985. It matures in 140 days. Plants are tall, seeds are white round, smooth, bold (210 g/1000 seed). It is resistant to powdery mildew. Average yield is 15-20 q/ha.

Rachna : Its pedigree is T 163 x T 10 and was identified in 1982. It matures in 125-130 days. Plants are erect, tall, flowers white, seeds creamy white, round, smooth, bold (210 g/1000 seed). It is resistant to powdery mildew. Average yield is 15-20 q/ha.

3.5.7 Agronomic management

3.5.7.1 Sowing parameters

The land for pea cultivation should be prepared well. It requires a well pulverized soil. One deep ploughing followed by two to three harrowing and leveling is sufficient to prepare the land to desired tilth for its production. The land should be free from volunteer plants and stubbles.

Cultivar should obtained nucleus / breeder / foundation seed from source approved by seed certification agency. Time of sowing is the most important factor influencing the yield of pea. Sowing of crop at optimum time and under optimum moisture conditions appears to be very important component of management. Sowing of pea should be done in mid October to November in plains, October to November and February to March in the hills.

The optimum sowing time of pea appears to be generally during the second half of October in North plains, November in parts of Bihar and West Bengal and first week of October in southern parts of Madhya Pradesh and Maharashtra. Planting in mid-October gave optimum results of Pantnagar (Saxena, 1977a). Planting at 30-50% of available soil moisture status gave best germination. Ram and Singh (1989) observed that maximum grain yield of pea was obtained on October 25 sowing.

The sowing of pea should be done in rows with seed drill or behind the plough at 3-4cm depth. A row spacing of 30-45 cm is generally recommended for sowing of pea depending upon the varieties and sowing time. Tall varieties of optimum sowing time should be sown at 45 cm row distance. With the delay in sowing, spacing should be reduced. In case of dwarf varieties, closer spacing is superior and more number of plants may be accommodated per unit area. The optimum spacing also differs depending upon the environmental conditions and plant type. The experiments conducted at different centres of Co-ordinated Pulses Improvement Project that 40-60 kg/ha seed is sufficient for small seeded varieties whereas for bold seeded ones, seed rate of 70-100 kg/ha should be used (DPR, 1983).

Before sowing, seed should be treated with mancozeb or carbendazim @ 2 g per kg of seed to control seed and soil-borne fungal diseases. After treated the pea seeds with fungicides, appropriate *Rhizobium* culture @ 5 g/kg of seeds is to be needed. Seed inoculation with *Rhizobium* culture for legumes which is grown first time in the field is necessary. *Rhizobium* inoculation may increase the seed yield and protein content of seed.

3.5.7.2 Nutrient management

Addition of manures and fertilizer increased the growth and grain yield of pea. Responses of pea to fertilizers are comparatively higher among pulses. Application of 5 t FYM/compost at the time of land preparation along with 20-25 kg N and 50-60 kg P_2O_5 and 20-25 kg K_2O/ha as basal at the time of sowing is required for pea seed production in medium to low fertile soil. On an average, field pea responds significantly upto 50 kg P_2O_5/ha at different centres Panwar (1977). Foliar application of 0.2% DAP (Diammonium phosphate) at branching and another at flower bud initiation has been found very effective for getting better seed yield of pea.

3.5.7.3 Water management

Irrigation in pea mainly depends on the amount and distribution of winter rains, residual moisture at the time of sowing and type of soil. Give two to three irrigations at an interval of 15-20 days, according to weather and soil conditions, before the flowering and fruiting period. Two irrigations at pre-flowering and flowering stages gave higher yields of pea in a season with scanty winter rains. At Dholi, field pea responded significantly to two irrigations. First irrigation at branching stage and second irrigation either at flowering stage or pod initiation was found to be optimal (Ali, 1989b). On the basis of mean yield data, it may be summarized that on sandy loam soils, one irrigation to pea at flower bud initiation stage is the most beneficial. In absence of winter rains, two irrigations, one each at branching stage and pod initiation stage appeared to be ideal. The mean response to one and two irrigations was 331 kg and 482 kg/ha, respectively.

3.5.7.4 Weed management

Pea is a slow growing crop at early stage and weeds faster occupy the space available in between the rows. However, spreading nature of plants cover major surface area after some time and weed problem does not continue like other pulses. One hand weeding 20-25 days after sowing (DAS) may be sufficient to keep to seed field clean of weeds. Two hoeing given at 30 and 45 days after sowing proved superior to one hoeing at either of these stages (Mani, 1972). Several herbicides are now also available which are effective in controlling the weeds in peas and have been tested in various centers. Few herbicides like Basalin, Lasso, Pendimethalin and MCPB are promising. Excellent control of weeds and increase in grain yield of pea by post-emergence application of 0.84 kg a.i./ha MCPB has been achieved at IARI (Ahlawat and Singh, 1979).

3.5.8 Plant protection measures

Powdery mildew: In India, this disease is a major threat to pea cultivation. The disease is caused by *Erysiphe polygonide* and appears in an epidemic form almost every year towards the end of January and in February when plants are in pod stage. The affected plants show white powdery mass on leaves, and later on stem and pod also. Dusting of sulphur (fine powder) at the rate of 30 kg/ha gives very good control. The disease can also be satisfactorily controlled by spraying fungicides such as dinocap and binapacryl. Early maturing varieties are more susceptible than late ones. Pea variety 'Rachna' and 'T-163' showed field resistance to the disease.

Rust : The disease is caused by fungus *Uromyces fabae* (Pers.) de Bray. The disease is widely prevalent in humid areas of Northern India. It appears as early as in December and continues to increase upto March. The affected plants show brown to dark brown pustules on leaves, stems and even pod also. The disease can, however, be controlled by spraying of Mancozeb thrice at the rate of 3 kg/ha. The first spraying should be given when the disease just appears. The other two sprays can follow at 14 days intervals.

Stem fly : Affected plants show yellowing and later they wilt. Crop may be protected with basal application of systemic insecticides such as phorate, disulfoton etc. as basal at the time of sowing in the soil.

Pod borer : This is very harmful insect damaging number of pulse crops. Damaging is more in late maturing crop. This pod borer is serious pest of pea to damage the pod and feeds on the seed. Spraying of 0.04% monocrotophos emulsion is effective in controlling this insect.

3.5.9 Roguing

For seed production, rouging must be done at least three times at vegetative stage, flowering stage and maturity stage. All the off-type plants distinguishably on the basis of plant characters must be removed before flowering to ensure pure seed production of pea. The diseased plants affected by pea mosaic, foot rot and blight should be removed from the seed field from time to time as required.

3.5.10 Harvesting and threshing

The pea crop for seed should be harvested when the seeds are fully mature. The harvesting may be started when ninety per cent of the pods turn brown. The plants may be uprooted or cut with sickles and stacked in small heaps in the field to dry in the sun for a few days (1 to 2 weeks). Then it should be shifted to threshing floors. The threshing may be done either by beating the plants with sticks or with the help of suitable thresher. Excessive beating should be avoided so that the seed is not injured. After threshing operation the seed is winnowed and cleaned. This should be dried to 9% moisture before storage. During harvesting and threshing operations, care should be taken to avoid mechanical and moistures (contact of the seed with moist soil).

3.5.11 Seed yield

With the use of improved variety and good management practices, seed yield of 20-25 q/ha may be obtained.

References

Ahlawat, I.P.S. and Singh, A. (1979). Progress Report of Agronomy, IARI, New Delhi.

Ali, Masood (1989b). Consolidated Report of Pulses. Agronomy, DPR, Kanpur (U.P.).

DPR (1983). Consolidated report on rabi pulses, DPR, Kanpur (U.P.)

Mani, V.S. (1972). Proc. Eighth Workshop (*Rabi* Pulses), ICAR : 158-161.

Ram Sri and Singh, N.P. (1989). Indian Journal of Pulses Research 2(1) : 68-72.

Saxena, M.C. (1977a). Expt. Stn. Tech. Bull. G. B. Pant Univ. Agric. & Tech., Pantnagar, India. 101: 1-77.

Vavilov, N.I. (1926). Studies on the origin of cultivated plants. Bulletin of applied Botany and Plant Breeding. 16(2): 50-53.

Panwar, K.S. (1977). Indian J. Agri Res. 11(1): 51-52.

3.6 Cowpea

Cowpea [*Vigna unguiculata* (L.) Walp] is known as Lobia. It is a most versatile *kharif* pulse in India because of its draught and heat tolerating characters, smothering nature, soil restoring properties and multi-purpose uses. Cowpea is one of the most important crops in the semi-arid tropics where they are grown mainly for their mature seeds, but also for their immature pods and leaves are used as human food. Cowpea can also be used for fodder and green manure purpose. Cowpea gives a heavy vegetative growth and covers the ground so well that it checks the soil erosion in problem areas. On an average, cowpea seeds contain 23% protein but there is much variation in protein content so that breeders may succeed in producing cultivars with seeds containing much more. It is a rich source of calcium and iron. Cowpea as a single season crop under ideal condition can fix nitrogen about 80-90 kg/ha. Cowpea can be grown in dry-land areas due to its fast growth, short duration and low water requirement characteristics.

3.6.1 Origin and distribution

According to Vavilov (1951), the cultivated cowpea originated in the Indian Subcontinent. However, evidences support, central Africa to be the centre of origin of cowpea as most of wild species have been noticed here (Faris 1965). It is considered that cowpea is spread to rest of the world from Ethiopia and West Africa. It is also grown in North and South America, Asia, central and southern parts of Europe and Australia. In India, cowpea is mostly grown in Gujarat, UP, Rajasthan, West Bengal, Tamil Nadu, Andhra Pradesh, Kerala and Odisha. Cowpea is usually used during auspicious ceremonies by the "Hindu" in India.

3.6.2 Botany

Cowpea belongs to family leguminosae. Cultivated cowpeas are very diverse, usually glabrous annual herbs, with a strong deep tap root and many branches in the surface soil. The root nodules are smooth and spherical, about 5 mm in diameter, they are numerous on the tap root and main branches, but sparse on the smaller roots. The terminal leaflet of the trifoliate leaves is commonly around 12 cm long (rarely as long as 16 cm). The stipules are large and spurred at the base, the stipules are inconspicuous. The large showy flowers are commonly white or purple, rarely yellow, with a standard petal 2-3 cm across. They occur in alternate pairs on a long axillary peduncle, from 2 cm to rarely more than 30 cm long. There are large extra floral nectaries between each pair of flowers. The stamens are diadelphoses and the ovary is straight with a style bent at right angles, hairy on the inner side and extended as a beak beyond the lateral stigma. The flowers are cleistogamous and although 1-2% crossing occurs. Cowpea cultivars are usually inbreeding, more or less homogenous pure lines. The pods of cultivars are indehiscent and vary greatly in colour and size. Seed germination is epigeal. The chromosome number is 2n = 22, 24.

Classification

Purseglove (1968) classified all cultivated types of cowpea into following three groups.

1. *Vigna unguiculata* subspecies *unguiculata*
2. *Vigna unguiculata* subspecies *cylindrical* (catjang)
3. *Vigna unguiculata* subspecies *sesquipedalis* (Yardlong, asparagus bean)

A distinct group of early maturing, more or less upright, day neutral forms of subspecies *unguiculata* probably evolved in the transition zone between the west African rain forest and Savanna, where rainfall is bimodally distributed and there is a marked dry period during the rainy season. The day neutral forms of sub-species *unguiculata* are the chief forage cultivars in the advanced agriculture of other parts of tropics. Other two sub-species *cylindrical* and *sesquipedalis* evolved through selection from sub-species *unguiculata* after its introduction in India. *Cylindrica* has determinate and upright plants and is chiefly grown for forage. *Sesquipedalis* has spreading habit and is mainly grown for long (30-100 cm) pods and is used as vegetables. Pod and seed size are the chief distinguishing features of three sub-species as follows.

(i) Sub-species *unguiculata* : Pods 10-30 cm long pendent; seeds 5-12 mm long, very rarely shorter than 6 mm.

(ii) Sub-species *cylindrical* : Pods 7.5-13 cm long, usually erect; seeds 5-6 mm long.

(iii) Sub-species *sesquipedalis* : Pods larger than 30 cm, flabby; seeds usually 8-12 mm long.

3.6.3 Climate

Cowpea or lobia can be grown in all tropical and subtropical climates. The best temperature for germination of cowpea seed is 12-15°C. Cowpea can thrives well between 21 and 35°C. The optimum temperature for primary and secondary root nodulation is 24°C and 33°C, respectively. Cowpea is a short-day plant and requires a minimum critical day length of 12.5 hours. For higher production of cowpea, 27°C day temperature and 22°C of night temperature is required. Area receiving 60 cm rainfall are suitable for short duration varieties. It can tolerate drought and therefore, is an important crop of dry areas. It is very susceptible to water-logging and is not suitable for heavy rainfall areas.

3.6.4 Agronomic management

3.6.4.1 Selection and preparation of land

Select lands on which same kind of crop was not sown in the previous season with the same variety of certified seed. Cowpea can be grown successfully on a variety of soils with proper drainage. Well drained loam to sandy loam soils is best for cowpea seed production. Soil should be neutral in pH. Saline and sodic soils are unfit for cowpea cultivation. Field should be prepared by giving one deep ploughing followed by two to three cross-harrowing and planking. In normal soils, primary roots of cowpea penetrate upto 50-60 cm depth. For summer crop give a pre-irrigation immediately after harvesting of *rabi* crop. When the field comes in proper condition land should be prepared by giving two or three harrowing. Land should be well leveled to avoid even temporary water-logging. The land should be free from volunteer plant.

3.6.4.2 Sowing parameters

Nucleus / breeder / foundation seeds obtained from an approved certification agency should be used for sowing of cowpea seed crop. Cowpea is sown generally with the onset of monsoon. Sowing of cowpea in mid June to mid July is recommended for higher seed yields. Late sown affects plant growth adversely which results in poor number of flowers and less grain and pod yield. In South India, *rabi* cowpea is also grown October and November. Spring crop in north India is sown in last fortnight of February or first fortnight of March. Early sowing in February results in poor germination due to low temperature and sowing late in March affects grain yield adversely due to forced ripening.

Under Bihar and West Bengal conditions where the rainfall is high, late rainy season sowing of cowpea (Late in July or early August) allows more grain and less vegetative growth.

Cowpea seed can be sown on flat, or raised beds, with a drill or dibbled in rows. The depth of seeding should be 2.5 cm to 3 cm. For seed production row to row spacing is 45-60 cm depending upon the duration of variety and time of sowing. 10-15 cm of plant spacing are maintained. For higher seed production optimum plant population of cowpea for one hectare is 4-6 lakhs. However, seed rate of cowpea varies according to the purpose it is grown. For grain or vegetable crops, 20-25 kg/ha seeds are required. Yadav (1992) reported that during summer season, vegetative growth is comparatively less. Therefore, closer sowing with higher seed rate should be followed. A row spacing of 30 cm and plant spacing of 7 cm have been found to be the optimum for cowpea. Seed rate should be increased upto 30 kg for grain production.

For the production of seed against soil-borne / seed-borne diseases, seed should be treated with fungicide like Thiram + Bavistin (1:1) @ 2.5 g/kg of seed. After treating the seed with those fungicides, appropriate *Rhizobium* culture is to be used. Cowpea seeds are inoculated with *Rhizobium* spp. (cowpea type) at the rate of 5 g culture per kg seed, regardless of seed size. Germination and seedling vigour are significantly influenced by seed size but not days to maturity. Small and medium seeds genotypes had better germination and mobilization efficiency than the bold seeded genotypes, whereas bold seeded genotypes had higher vigor index.

3.6.4.3 Isolation requirements

Agrawal (1995) reported that cowpea is largely a self-pollinated crop but some cross-pollination by insects has been recorded. In areas of abundant insect activity, varieties should be isolated at least 130 m apart. Indian minimum seed certification standards require that a seed crop be isolated, at least by 50 cm for foundation seed class and 25 m for certified class, from fields of other varieties, or of same variety not conforming to varietal purity requirements of certification.

3.6.4.4 Nutrient management

Cowpea needs a small quantity of nitrogen for early growth period like other pulses, particularly on those soils which are poor in organic matter. For seed production of cowpea application of FYM @ 1-2 t/ha at the time of land preparation is effective. In sandy loam soils, application of FYM @ 4-5 t/ha increases the seed yield of cowpea. The recommended doses of nitrogen, phosphorus and potassium are 20-50-30 kg N, P_2O_5 and K_2O/ha respectively. All fertilizers should be applied as basal by drilling 5-6 cm below the seed. Rathore (2002) reported that highest seed yield of cowpea with 50 kg P_2O_5/ha applied to the soil at sowing time and 25 kg P_2O_5/ha foliar application at the first flush of flowering. Phosphorus should be applied through DAP. Ali (1989b) reported significant increase in the yield of *rabi* cowpea after *kharif* rice at Bangalore and the highest seed yield was recorded with application fo 18 kg N + 46 kg P_2O_5/ha. Application of 20 kg K_2O/ha also increased the yield. All the nutrients and culture alone and in combinations increased the yield significantly.

3.6.4.5 Water management

In rainy season cowpea, proper care should be taken to drain water from the field. Normally no irrigation is needed in *kharif* season crop, unless there is prolonged drought due to absence of rains. Cowpea is a deep-rooted pulse crop with tap root system. In *kharif* season, it is grown with monsoon rains and utilizes the rain moisture effectively. Both summer and spring cowpea can be grown only with irrigation facilities. Early sown rainy season cowpea needs one or two pre-sowing irrigations (70 mm water for irrigation). In summer, number and frequency of irrigations depend upon the soil type and prevailing weather. Two irrigations are optimal after a pre-sowing irrigation. One irrigation should be given at flowering stage (30-30 DAS) and the second at grain development stage (42-45 DAS). Water requirement of cowpea is 300-400 mm. The best criteria for irrigation in cowpea is the range of soil moisture between medium dry to wet, crop should be irrigated at 25-50% depletion of available water from its maximum root depth (Rathore, 2002).

3.6.4.6 Weed management

Grain legumes suffer from weed competition due to their initial slow growth. Keep the seed field free of weeds. One to two weeding/hoeing in early stages of crop growth may be required. Crop yield losses due to weeds competition in cowpea have been estimated to range from 30-50%. Cowpea has a smothering effect on weeds. Because of fast and early growth, it attains full ground cover stage at 30 DAS. Thus the crop needs protection from weeds for initial 15-20 DAS. Two weedings at 15 and 30 days of crop growth ensure an effective weed control. During rainy season, weeds can be controlled by the use of chemicals trifluralin @ 1 kg/ha as pre-emergence application (Rathore, 2002).

3.6.4.7 Roguing

To maintain the seed purity, the off-type plants and diseased plants affected by blight, cowpea mosaic virus and anthracnose should be removed from the field from time to time as required. So, for seed crop rouging is done at the following three stages.

(i) Pre-flowering stage-depending on leaf and stem character

(ii) After flowering-depending on inflorescence character

(iii) At maturity-depending on grain character

3.6.4.8 Plant protection measures

Cowpea as a grain legume being rich in protein may be attacked by a number of pests (*viz.* stem boring maggots, aphids, jassids and white flies), and various fungal, bacterial and viral diseases. These pests may be controlled by spraying the crop with 0.1% metasystox or 0.4% monocrotophos 40 EC; which may be repeated 15-20 days intervals if necessary. The disease *Cercospora* leaf spot and Anthracnose can be controlled by spraying with copper oxychloride at 1.5 to 2.5 kg in 625 litre of water/ha as soon as initial symptoms of the disease appear and repeat at 15 days interval. Collect the crop residues from the field and destroy it. Bacterial blight and canker are seed-borne disease and can be controlled by seed treatment with streptomycine sulphate (500 ppm) for 30 minutes and spraying the crop with 0.2% carboxin when disease appears. Rust is a soil and seed-borne disease and can be controlled by spraying the crop with mancozeb @ 2 kg/ha in 1000 litres of water. Cowpea mosaic virus is transmitted by aphids (*Aphis gossypii*) and seed to seed transmissible. This can be controlled by use of healthy seeds and spray 0.1% metasystox for aphid control.

3.6.4.9 Variety

Different improved varieties of cowpea have been recommended according to their availability, and adaptability of different agro-climatic conditions. However, same variety may be grown for more than one purpose like grain fodder and vegetable. The varieties recommended for use as pulse grain are C-152, FS-68, Cowpea 74, CO-3 etc., for use as fodder are type-2, Pusa Rituraj, Pusa Barsati etc. and for use as vegetable are Pusa Phalguni, Pusa dofasli, Pusa barsati and FS-68. Most of grain varieties showed shy growth habit, late maturing and susceptibility to viral and bacterial diseases. Now most of the breeding programme throughout country is developing short duration, drought resistant, high yielding and disease resistant varieties. The following varieties of cowpea have been recommended for different zones (Rathore, 2002).

Zone	States	Name of Varieties
North-western zone	Punjab. Haryana, Western UP, Rajasthan, Himachal Pradesh, Jammu and Kashmir	Pusa 152, T5269, F568, RS9,V-16 (Amba), Cowpea 74, JC 5
North-eastern Zone	Eastern UP, Bihar, West Bengal, Odisha, Assam, Maharashtra	Pusa 152, T 5269, V-16, RC 19, FGS-1, Gomti, Type 2
Central Zone	Madhya Pradesh, Gujarat, Maharashtra	Pusa 152, V-16, V-240, K-11, K-14, Gujarat cowpea 1, Gujarat cowpea 2
Peninsular zone	Andhra Pradesh, Karnataka, Tamil Nadu	Pusa 152, V-16, CO 1, CO2, CO 4, Krishnamani

The important characters of promising varieties of cowpea are given below:

1.	**C-152:** Developed by selection from Iranian variety P-42 at IARI; matures in 105-110 days, suitable for *kharif* season planting in Punjab, Haryana, Delhi, U.P., Bihar, M.P. and Maharashtra, yields about 15-20 quintals grain/ha; medium bold brown seed. Good for fodder, for pure and mixed cropping.
2.	**Pusa phalguni:** A selection from 'Doloque on Tonkin' from Canada. It is a busy dwarf variety (50-60 cm plant height), best suited for sowing in Feb.-March in northern plains, matures in 60-65 days and yields 10-12 q/ha grains and 50-60 q/ha green pods.
3.	**Pusa Dofasli:** A cross between Pusa phalguni x Philippines selection. It is an improvement over Pusa phalguni for pod quality. Suitable for growing in *kharif* and summer season in northern plains. The Pod is about 18 cm long and gives the first picking about a week later than Pusa phalguni. It is photo-insensitive variety and matures in 60-65 days and produces 30-40 q pods/ha and 10-12 q grain/ha.
4.	**Pusa Rituraj:** It is a recent selection, developed by NBPGR, New Delhi. A highly photo-insensitive variety grown both in summer and *kharif* season in most parts of north India. The plant is bushy in nature and matures in 60-65 days. It yields pods so profusely, as to cover foliage appearing like a fan. Average green pod yield is about 80-85 q/ha.
5.	**FS – 68:** It is sown both during rainy and spring season. It takes about a week later than Pusa dofasli in maturity, suitable for both grain and vegetable, pods small and green; seeds small and white; yield 10-12 q/ha.
6.	**CO 3 :** Developed by selection from C-152. It is a semi-erect variety, matures in 80-85 days and yields 10-15 q grain/ha.
7.	**Cowpea-74:** Suitable for growing in Punjab and Haryana both for grain and fodder purpose. It is resistant to viral diseases. It yields 300 q/ha green fodder and 12-15 q grain/ha.
8.	**K 11 :** It was developed by selection in Madhya Pradesh. It matures in 90 days. Seeds are white in colour. The average yield is 15-20 q/ha.
9.	**RC-10 :** A cross between Virginia x New Era. Best suited for growing in NE plain zone and Rajasthan. It matures in 60-65 days and yields 10-12 q grains/ha.
10.	**Type-2 :** Developed by selection in 1955 from a material collected from Jhansi. Matures in 125-130 days, spreading type, good for grain, fodder and green manure. Average yield is 15-20 q/ha.

11.	**Pusa barsati:** It is selection seed material imported from Philippines. Earliness is the special characteristic of this variety. It is suitable for growing in monsoon season in Punjab, Haryana, U.P., Delhi and M.P. It matures in 75-80 days and yields 85-90 q green pods/ha.
12.	**CO 1:** Developed by selection in 1970 from local germplasm. Good for grains and fodder purpose in Tamil Nadu. It is a semi-spreading bold seeded variety which matures in 100-105 days. Average grain yield is 5-6 q/ha.
13.	**CO 2:** It was developed by the cross of C 152 x C 419 in 1978. It matures in 85 days. Plants are spreading type, seeds are kidney shaped, reddish brown and mottled.
14.	**CO 4:** It is a selection from Russian Gaint fodder variety and was released in 1983 for irrigated and dryland conditions; Plants are bushy erect with green foliage. Seeds are grayish brown and medium bold (175 g/1000 seed).
15.	**Gomti (5286-3)** : It is a selection from local germplasm, developed in 1974. It matures in 80 days. Plants are spreading with long greenish white pods having 14-15 seeds/pod. Average yield is 15-18 q/ha. It is suitable for summer and *kharif* planting in Chhotanagpur region.
16.	**JC 5** : It was also developed by selection from local germplasm in 1955. Its maturity period is 80-90 days. It is suitable for cultivation in arid and semi-arid tracts of Rajasthan; average yield 5-7 q/ha.
17	**V 240** : It is a mutant of Pusa Phalguni, developed in 1984 and is suitable for general cultivation in western U.P. Its maturity period is 70 days. Seeds are red coloured and round shaped. Average seed yield is 8-10 q/ha.
18	**RS 9** : It was developed by the selection from local germplasm in 1955. Its duration is 90-95 days. Seeds are small and buff coloured. Average yield is 6-8 q/ha.

3.6.4.10 Harvesting and threshing

The seed crop may be harvested or the pods may be picked when they are fully dried. Crops that mature in relatively dry months from November to May, do not pose problems of harvesting, drying and processing. But under multiple cropping harvesting, post-harvesting operations like drying and processing create problem, when the crops are to be harvested in wet months.

In vigna group of grain legumes, because non-synchronous maturity of the crop hand picking of the matured crops, their drying and processing become too expensive, particularly when they are to be harvested in wet months. The late August of early September sown *Vigna* group of grain legumes, maturing (when 80-90% of the pods have become brown) after the cessation of monsoon rains can be uprooted or cut at ground level and then can be transported and dried in thin layers with occasional turnings and can be threshed by bullock or other power-driven equipments. Threshing or beating by sticks to bring out the grains in *Vigna* group of grain legumes cause a lot of damages and bruises on the seed coat which are not desirable. During threshing operation care should be taken to avoid mechanical admixtures, contact with moist soil and injury to seeds.

3.6.4.11 Seed yield

Average seed yield of cowpea varies from 15 to 20 q/ha and 55-65 quintals of straw per hectare. Vegetable cowpea produces 50-60 quintals of green pods per hectare, while fodder crop yields about 300-400 quintals of green fodder per hectare.

3.6.4.12 Seed quality

Seeds of grain legumes which are normally self-pollinated, can be sown theoretically for several generations without genetic deterioration. In practice, however, progressive deterioration of the original stocks occurs due to mixture with other varieties, etc. Therefore, every 4-6 years new seed, produced under strict control, is required to maintain the yielding potential of an improved variety. Availability of quality seeds of improved cultivars is considered crucial for realizing productivity and adoption of cultivars in different agro-climatic conditions. Genetic purity, uniformity should be confirmed to standards of particular cultivar.

Nutritive composition of cowpea seed per 100 g of edible portion

Composition	Amount
Water	11%
Calories per 100 g	342
Protein	23.4%
Fat	1.8%
Carbohydrate	60.3%
Calcium	76 mg/100 g
Iron	5.7 mg/100 g
Thiamine	0.92%
Riboflavin	0.18 mg/100 g
Niacin	1.9 mg/100 g

3.6.4.13 Storage

The amount of moisture in the seed is probably the most important factor influencing the length of time the seeds remain viable. Cowpea seeds for storage should contain less than 12% moisture. Through sun-drying the moisture content in seed can be reduced to 9%. Any seed at the stage of physiological maturity has maximum germination and vigour (dormancy factor can be ignored). After this stage, the germination and vigour decline. However, the rate of decline depends upon the weather conditions prevailing during ripening of the seed and post-harvest storage condition. Before storage the seeds can be treated with captan or thiram to protect them against anthracnose (*Colletotrichum lindomuthianum*), wilt (*Fusarium oxysporum*) and damping-off disease (*Pythium* spp.). Streptomycin sulphate controls bacterial blight (*Xanthomonas vignicola*). The best way to combat seed-borne disease of pulse crop is to use only healthy seed from a clean crop. Studies conducted in India observed that seeds of pulse treated with malathion and stores in polythene bags retain their viability over a period of 8 months. The storage conditions should be checked at regular intervals for cleanliness, presence of insects, temperature, relative humidity and moisture content of the seeds.

References

Ali, Masood (1989). Consolidated Rep. of rabi pulse, Agronomy DPR; Kanpur (UP).

Aggrawal, R.L. (1995). Seed Technology. Oxford and IBH Publishing Co., New Delhi, 2nd Edition. P. 145.

Faris, D.G. (1965). Cam. J. Genet. Cytol. 7: 433-452.

Rathore, R. (2002). Techniques and management of field crop production. Published by Agrobios (India): 308-316.

Vavilov (1951). Chramica Prot. 13 : 364.

Yadav, D.S. (1992). Pulse crops (Production Technology), Kalyani Publishers, New Delhi : 162-171.

3.7 Lathyrus / Khesari

Khesari (*Lathyrus sativus* L.) also known as grasspea, chickling pea, khesari, is winter season crop and used as 'dal' in human consumption. It is also used as feed and green fodder by the animals. Khesari is most resistant crop and can be grown in adverse situation. It is resistant to moisture stress as well as high moisture condition.

Poor people use the grain as pulse, particularly in times of scarcity. The unripe pods are occasionally used as a green vegetable. The grain and plant parts contain a neurotoxin, BOAA, which has been shown to be the inciting agent of Lathyrism, a paralysis of lower limbs, endemic in some regions of Madhya Pradesh, Bihar and Uttar Pradesh. Ill effects are noticed when lathyrus forms a significant part of the diet over a considerable period. Steeping the pulse over-night in water and reject the water effectively detoxifies the pulse. Parboiling has also be the same effect. The variety 'Pusa 24' and 'Nirmal' are available which has a very low BOAA content and therefore, is not likely to cause lathyrism.

Khesari is chiefly grown on the residual moisture in rice fields as utera crops. More than 60% of the area under Lathyrus is established under *utera* or *paira* condition. *Utera* (*Paira*) cultivation refer to the sowing of seeds in standing paddy crops 15-20 days prior to its harvest, when soil moisture is favourable for sowing or in other words, it is a sort of relay cropping system. This is usually done under conditions where late *aman* paddy is harvested by the end of November or early December on low land heavy soils. The soil is not immediately good for ploughing due to high soil moisture condition. Late sowing of crop may not be good for productivity and the soil moisture may limit a growth late in the season and the land vacated by rice has usually no irrigation resource.

3.7.1 Origin and distribution

Lathyrus is indigenous to Europe and West Asia. It is mainly grown in the Indo-Gangetic Plains comprising of Bihar, Madhya Pradesh, Uttar Pradesh, Odisha, Assam, West Bengal, Middle Gujarat and some parts of Deccan.

3.7.2 Botany

Grass pea is an herbaceous annual with slender, glabrous, freely branched winged procumbent stems; tap root well developed. Leaves alternate, pinnate terminating in 3-5 tendrils; prominent, leaf like stipules, ovate, entire, attached to petiole at base; leaflets 2-4 entire, linear, lanceolate, sessile, up 7.5 x 1.3 cm. Flowers solitary in leaf axils, 1.5 cm long; peduncle long; corolla blue or purple with white keel, stamens diadelphos, anther uniform, stigma spatulate. Pods flattened, oblong, upto 4cm long, two-winged dorsally, upto five seeded. Seeds wedge shaped, angular, colour range white brown, sometimes mottled. Seed germination is hypogeal.

Classification

This is a fairly large genus which includes perennial and annual herbs of Northern hemisphere temperate regions principally. The stems climb by tendrils rather than by twining; the stems are winged or angular; leaves are pinnate often with a terminal tendril; flowers in axillary recims or solitary. This genus includes some species grown for forage (the velchling), some as ornamentals, *Lathyrus odoratus* L. (Sweet pea) and *Lathyrus latifolius* L. (Everlasting pea) and one species *Lathyrus sativus* as a pulse. Chromosome number is 2n = 14.

3.7.3 Climate

Khesari is a winter season pulse crop. It requires a cool climate. Khesari is most resistant crop and can be grown in adverse situation. It is not affected either by excess of water in the field at the time of sowing or by shortage of water during the growing season. Its cultivation is very common in the late paddy areas. It is grown either as sole crop of the year or as a second crop after paddy, being often sown as an utera crop in standing paddy.

3.7.4 Variety

Pusa 24 : Developed by selection in 1976 at IARI, Delhi, matures in 150-160 days, seeds reddish brown, neuro-toxin content is very low. It has wide adaptability and is suitable for major khesari growing areas.

B-1 (Nirmal) : Selection at Berhampur, Pulse and Oilseed Research Station; Neurotoxin content is low (0.2%).

BIOL-212 : Developed from Pulse and Oilseed Research Station, Berhampur for north-east plains and central zone to replace old varieties; high yield potential with low neurotoxin content.

3.7.5 Agronomic management

3.7.5.1 Selection of land

Field selected for the seed production of lathyrus should not have been used for the same crops in the previous year, unless the same variety was grown using a certified seed. It is generally grown on clay loam or clay soils which remain till long after its growth season is over. Deep black soils produce excellent crops. The crop is not injured much by the cracking of black soils. Being hardy, it can be grown on lands which are suited to no other crops.

Two or three ploughings are given soon after rains to prepare a fairly clean seedbed. On lands submerged under water till late October in November, it is sown by broadcast on the muddy soils, where it sprout and grows in spite of the subsequent hardening of the soil almost to the consistence of stone, which will hardly sustain the more delicate crop of peas. Three to four medium tillage is practiced in *paira* cropping. More than 6% of the area under lathyrus is established under *utera* or *paira* condition.

Lathyrus is generally sown in second fortnight of October to First fortnight of November and Mid October to Mid November in west when the duration of crop is less. It can be sown little bit later period provided there is irrigation facilities.

Under normal condition, it should be done in rows at a distance of about 30 cm for row to row and 10 cm for plant to plant. The seed rate is 40-50 kg/ha for broadcasting and 25-30 kg/ha for line sowing. For *paira* cropping, 30-40% extra seeds should be provided to encounter the possible plant damage. Under adverse conditions such as low soil moisture, seed rate may be increased by 10% for germination and getting optimum plant population. Cultivation is done with nucleus / breeder / foundation seeds from a source approved by seed certification agency.

3.7.5.2 Seed treatment

Khesari seed should be treated with fungicide like mancozeb @ 2.5 g per kg of seed to protect from seed-borne disease. After treating the seed with fungicide, appropriate *Rhizobium* culture i.e. (*Rhizobium leguminosarum*) is to be used for better yield. Seed inoculation with *Rhizobium* culture for legumes which is grown first time in the field is necessary.

3.7.5.3 Isolation requirement

An isolation of three meters from other lathyrus fields is sufficient for pure seed production.

3.7.5.4 Nutrient management

Apply FYM @ 3-4 t/ha along with 20 kg N, 40-50 kg P_2O_5 and K_2O/ha as basal during the time of land preparation. Addition of 20 kg DAP (Diammonium phosphate) at pre-flowering or early pod filling stage is effective for higher seed yield. In rainfed area, maximum seed yield of lathyrus was recorded in rice-lathyrus (as *utera* crop)-sesame when rice and sesame in the sequence received 50% recommended dose of N, P and K in addition with 10 t FYM/ha and green manuring to rice (Ghosh and Mondal, 2002; Mondal *et al.,* 2008).

3.7.5.5 Water management

Traditionally lathyrus is not irrigated. Application of irrigation to maintain soil moisture content about 80% of field capacity from time of emergence to maturity gave highest seed yield of 3.12 t/ha and fresh fodder yield of 34.5 t/ ha and 7.7 t/ha of hay. Irrigation at 30-35 days after sowing is recommended for better growth and seed yield.

3.7.5.6 Weed management

This crop cover the entire field quickly and normally checks the growth of weeds. One hand weeding at 25-30 days after sowing is sufficient.

3.7.5.7 Roguing

Off-type plants, diseased plants affected by blight, and weed plants should be rogued-out from the seed field from time to time as required for getting quality seeds.

3.7.5.8 Harvesting and threshing

Duration of crop is 130-140 days. The crop when ready to harvest pods becomes hard and yellowish in colour. The crop is cut with sickle, dried on the threshing floor for a week and threshed by biting with the stick or by thresher followed by winnowing. Then after sun-drying reduces the seed moisture upto 9% for proper storage.

3.7.5.9 Yield

Grain yield obtained 4-5 q/ha under rainfed condition, 10-15 q/ha under irrigated condition and 25-30 q/ha under all favourable conditions.

References

Ghosh, A. and Mondal, S.S. (2002). Performance of lathyrus (Lathyrus sativus) as utera crop in relation to growth duration of rice (*Oryza sativa*) under rainfed condition. In: Extended Summaries of Second International Agronomy Congress on Balancing Food and Environmental Security- A Continuing Challenge. 26-30 Novemebr, 2002, New Delhi, pp. 1079-1080.

Hukeri, S.B. and Sharma, A.K. (1980). Indian J. Agric. Sci. 50(2): 157-160.

Mondal, S.S. Jena, P. and Chandra, P. (2008). Integrated nutrient management for sustaining productivity and soil fertility build up in rice (*Oryza sativa*) – Lathyrus (*Lathyrus sativus*)– Sesame (Sesamum indicum) Cropping system. Indian J. agric. Sci 78(6): 534-536.

3.8 French bean

3.8.1 Origin and distribution

French bean (*Phaseolus vulgaris* L.) was domesticated in Central and South American more than 6,000 years ago. Wild forms with small black seeds are now to be found in tropical America. It was first cultivated with maize and it seems likely that two crops evolved together in a cereal/legume farming system in much the same way as cowpea, sorghum and pearlmillet in West Africa and Asia by the Spanish and Portuguese and are now grown through the cooler tropics but not in hot semi-arid or wet humid regions.

French bean is widely cultivated in the tropics, sub-tropics and temperate regions. It has spread to Europe, Africa and other parts of the world. In India, French is mainly grown hilly Himalayan region and Maharashtra. It has a great potential in irrigated areas of Northern Plains. Now, the varieties are also cultivated in north-eastern plains during *rabi* season under irrigated condition.

3.8.2 Botany

The French bean plants are bushy as well as climbing in nature. The bushy cultivars are day-neutral, early maturing dwarf plants, 20-60 cm tall with lateral and terminal inflorescence and determinate growth habit. Climbing cultivars are indeterminate, 2-3 tall to climb by support with twining. The pods are slender, 10-20 cm long, straight or curved and terminated by a pointed beak. The pod contains 4-6 seeds which vary greatly in size and colour. Germination is epigeal.

3.8.3 Climate

French bean is highly sensitive to frost and water-logging. The temperature range for proper growth of this crop is 10-27°C. Above 30°C, the flower drop is a serious problem. Similarly below 5 °C, the flowers and developing pods and branches are damaged.

3.8.4 Land requirement

For seed production of French bean fields on which the same crop was not grown in the previous season unless the crop in the previous season was of the same certified variety. Land should be high and well drained. Medium type of soils are most suitable for seed production of French bean. Soil should be free from excessive salts with near neutral in reaction.

3.8.5 Isolation requirements

Bean flowers are generally self-pollinated. Natural crossing, however, is more common than peas and the extent of cross-pollination may vary from 2-8%. The seed field must be isolated from fields of other varieties and the same variety not confirming to varietal purity requirements by at least 50 meters for foundation seed production and 25m for certified seed production.

3.8.6 Agronomic management

3.8.6.1 Sowing parameters

3.8.6.2 Soil and land preparation

French bean can be grown in light loamy to heavy clay soils under adequate soil moisture. Field should be prepared by one ploughing followed by two to three harrowing and leveling. Fine tilth is needed for French bean cultivation. The field should be free from stubbles and volunteer plants.

Cultivars should obtain nucleus/breeder/foundation seed from source approved by seed certification agency. Sowing of French bean in plains in January to February as *rabi* crop as well as August to September as *kharif* crop. In Hills, it is generally grown in the month of March to May.

Seed should be treated with fungicide like mancozeb @ 3g/kg of seed. After treating the seeds with fungicides, appropriate *Rhizobium* culture is to be used. *Rhizobium* inoculation may increase seed yield of the crop. Seed rate of 25-30 kg/ha for pole varieties and 85-90 kg/ha for bush varieties are needed.

French bean seeds are sown by drill or dibbled in rows. The depth of seeding should be 2 to 2.5 cm. For seed production row to row spacing is 60 to 75 cm apart and plant to plant distance is 10 to 15 cm. Sufficient moisture in the soil is essential for optimum germination. Keep the fields free from weeds. During growing periods two to three hoeing should be given for better condition of the seed crop.

3.8.6.3 Nutrient management

The crop is generally cultivated in marginal and sub-marginal lands of poor initial soil fertility. Hence, the care should be taken that this seed crop does not suffer from nutrient deficiency. Apply 10 t FYM/ha at the time of land preparation. The recommended doses of nitrogen, phosphorus and potassium are 20-50-30 kg N, P_2O_5 and K_2O/ha, respectively. The nitrogen, phosphorus and potassium fertilizers should be applied at the time of land preparation before sowing.

3.8.6.4 Irrigation management

One pre-sowing irrigation should be given if moisture in soil is not sufficient for germination of seed. One irrigation at flowering and another irrigation at pod development stage. It is essential to maintain available soil moisture above 50% during this stages gave better seed yield of French bean.

3.8.7 Plant protection measures (Control of pest and diseases)

Bean aphid and thrips may be controlled with a spray of 625 ml metasystox in 625 litre of water per hectare, which may be repeated at 15-20 days intervals if needed. Bean beetle may be controlled by spraying Monocrotophos @ 0.015%.

Anthracnose disease may be controlled by use of seed treated with fungicide and spray crop with copper fungicides. Powdery mildew may be controlled by dusting with sulphur or sulphur lime. Application should be started at the first appearance of the disease and repeated every 10-15 days, if needed. Rust may be controlled by dusting finely ground sulphur at 25-30 kg/ha before infection.

3.8.8 Variety

The improved varieties of French bean for cultivation in different zones are given below.

Improved Varieties of French bean

Variety	Source	Year of Release or notification	Yield (q/ha)	Days to maturity	Area of adoption	Characteristics
Him 1	HPKVV	1978	15-20	75-80	Himachal Pradesh	Seed light pink and red
VL-63	VPKAS	1982	12.0	73	Uttar Pradesh hills	Seed light red with deep patches
Udat (PDR 14)	IIPR	1987	18.0	125	NEPZ	Seed red-white variegated
Malaviya Rajmash 15	BHU	1989	15.0	120	NEPZ	Seed white
Malaviya Rajmash 137	BHU	1991	18.0	120	NEPZ	Seed red
HPR 35	HPKVV	1992	15.0	73	Maharastra	Seed red with purple red streaks
Amber (IPR 96-4)	IIPR	2002	15-16	139	NEPZ	Resistant to bean mosaic virus, seeds red

Source : Handbook of Agriculture, (2006), ICAR

3.8.9 Roguing

Roguing of seed crop is carefully done to maintain genetic purity. On the basis of foliage colour, plant type, flower and pod characteristics etc., the plant should be carefully checked and off-types removed from the field at pre-flowering, flowering and maturity stage. In addition to off-types, the plants affected by diseases namely bacterial blight, Anthracnose, Ascochyta blight and bean mosaic should also be removed from the fields from time to time as required.

3.8.10 Harvesting and threshing

For seed production, harvesting of French bean crop is done when a large percentage of pods are fully ripe and most of the remaining have turned yellow. Harvesting should be started before the lower pods become dry enough to shatter. The plants may be pulled by hand or cut with sickles. The crop after harvesting, is left in the field to dry for about 7-10 days. Later, it is threshed by bullocks or by threshers. While harvesting, care should be taken to keep the mechanical injury to a minimum. After threshing and cleaning the seed should be dried to moisture content below 9% before storage.

3.8.11 Seed yield

The average seed yield of French bean varies from 15 to 20 q/ha.

4

Seed Production of Oilseeds

4.1 Mustard

The principle oil yielding rape seed and mustard sp. cultivated in India are

1. Yellow sarson - *Brassica campestris* var. *yellow sarson*
2. Brown sarson - *Brassica campestris* var. *brown sarson*
3. Toria - *Brassica campestris* var. *toria*
4. Rye (True Indian mustard) - *B. juncea*
5. Banarasi Rye - *B. nigra*
6. Pahadi Rye - *B. juncea* var. *rugosa*

4.1.1 Land Requirement

Select seed field on which the *Brassica* sp. was not grown in the previous season unless the crop in the previous season was of the same variety and was grown from certified seed. To avoid disease control it is necessary to fallow a 3-4 year crop rotation so that a given field is selected one in every 3-4 years. In addition select field should be well drained, leveled and suitable for Brassica cultivation.

4.1.2 Isolation requirement

Indian minimum seed certification standards prescribe a minimum isolation distance of 100 m and 50 m for foundation seed class and 50 m and 25 m for certified seed class for self incompatible and self compatible type of crops from fields of other varieties of the same sp and the field of the same var not conforming to varietal purity requirement of certification. Varying amount of self & cross pollination, and self & cross incompatibility occurs in different species and different strains.

4.1.3 Agronomic Management

4.1.3.1 Preparation of land

Usually 1 ploughing followed by 3-4 harrowing and leveling are sufficient to prepare good tilth for rapeseed and mustard cultivation.

4.1.3.2 Sowing parameters

For toria ideal time for sowing mid October – 1st week of November. This is also for sarsons. Rye – ideal sowing time 1st week of September – mid October.

Obtain nuclear, foundation and certified seed from a source approved by the certifying agency.

Seed crop should be sown in rows with a depth of 1.25 – 1.5 cm.

Row – Row spacing 30 cm apart and Plant-Plant spacing should be 5-10 cm depending upon the architect of the plant. It has been found that row-row spacing of 40 cm is more advantage in our condition so for as the seed production of mustard is concerned.

Seed rate varies from 6-8 kg/ha.

4.1.3.3 Nutrient management

Rapeseed & mustard crop exhaust varying amount of nutrient from the soil in a heavy amount within a short time at a faster rate. Mustard crop yielding 4000 kg/ha of seed is found to remove 134 kg of N, 54 kg P_2O_5 and 72 kg K_2O. In general the N requirement of mustard varieties is higher than the rapeseed.

Table 4.1: Recommended doses (kg/ha) of fertilizer for mustard seed production

Fertilizer	Toria	Sarson	Rye
N	60	80	80-100
P_2O_5	30	40	40-50
K_2O	30	40	40-50

The whole amount of P_2O_5, K_2O and half of N should be applied as basal. The remaining half N should be applied at the time of 1st irrigation. Sulphur is called the 4th major nutrient after NPK specially for the Cruciferus crop. 2-3 times more S is required by the crop over cereals. About 11-12 kg S is required to produce 1 ton of rapeseed and mustard. Seeds of mustard contains 1.1% S as compared to 0.3% in pulses and 0.2% in cereals.

4.1.3.4 Water management

One pre-sowing irrigation is required to obtain uniform germination of the seeds. Each irrigation should be given 3-5 cm depth of water. After establishment of the crop first irrigation should be given at branching followed by flowering and pod development stages.

4.1.3.5 Intercultural operation

One hand weeding followed by thinning of the plant should be done when the young plant are 15-20 cm in height. Second hand weeding if required should be done at 35-40 DAS.

4.1.4 Plant Protection

For controlling mustard saw fly apply Chlorpyriphos @ 3 ml/lit of water, for the Aphid spraying of 1.5 lit of Metasystox 25 EC or 250-300 ml Demecron 100 EC dissolved in 1000 lit of water/ha and for controlling *Alternaria* blight spraying of Dithane M 45 @ 2-2.5 g/lit of water should be applied.

4.1.5 Roguing

All the off-type plant easily distinguishably on the basis of plant characteristics and other species of plant must be removed before flowering to ensure pure seed production. Remaining off type if any distinguishable on the basis of siliqua characteristics should be removed before maturity. *Argemone mexicana* is the most objectionable weed in rapeseed and mustard seed production, must be removed altogether as often as required.

4.1.6 Harvesting and Threshing

When the plants as well as siliqua turned into yellow colour and the seeds inside the siliqua turned into deep yellow or brown colour is right time for harvesting mustard. After harvesting the crop should be left in the field for 2-3 days and then bundled and taken into the well cleaned threshing floor for further processing. Threshing can be done either bullock or by beating with sticks. Before storage the seed should be sundried slowly to reduce the moisture content of the seed to 8%. Seed should be stored with proper labeling in a gunny bag and placed in well protected godown for future use.

4.1.7 Seed Yield

With good management practices yield of rapeseed and mustard should be around 15-20 q/ha.

Table 4.2: Improved varieties of rapeseed - mustard

State	Varieties
Kashmir valley Himachal Pradesh	Mustard: RCC4 Kiran, Pusa Swamia (IGC01) Gobhi Sarson: Nilam, Seetal, GSL1, GSL 2, Hyola 401 (Hybrid)
Jammu region, Panjab, Haryana, Delhi and Northern parts of Rajasthan	Toria: TL 15, PBT 37 Mustard: RH 30, Laxmi, RLM 619, CS 52, Geeta Kiran, Pusa Swamia (IGC01) Gobhi Sarson: GSL1, GSL 2, PGSH 51, Hyola 401 (Hybrid)
Western Rajasthan, Uttaranchal, Uttar Pradesh, Northern Madhya Pradesh	Toria: T9, Bhawani, PT 303, PT 507, JT1 Yellow sarson: Ragini Mustard: Jagannat, Baruna, Rohini, Urbashi, Basanti, Narendra Rai, Narendra Ageti, Rai 4, JM1 Kiran, Pusa Swamia (IGC01)
Gujrat, South Rajasthan, Maharastra West Bengal, Odisha, Jharkhand, Chattishgarh, Assam, Bihar, Manipur	Yellow sarson: Ragini Mustard: Pusa Jaikishan, GM 2, PCR 7 Toria: M 27, Agrani, Parvati, Anuradha; Yellow sarson: Binoy, Jhumka; Ragini Mustard: Pusa Bold, Pusa Agrani, Pusa bahar; Black mustard: Surya (LBM 428)
Karnataka, Andhra Pradesh, Tamilnadu	Mustard: Pusa Agrani (Sej 2), JM 1, GM 2, Bardan, Baruna

Recently released varieties / hybrids of mustard

Variety	Duration (days)	Yield (t/ha)	Area of adoption	Salient features
RH 761	141	2.86	Jammu, Punjab, Haryana, Delhi and Northern Rajasthan	Suitable for timely sown and rainfed conditions in *rabi* season, oil content 40.4%, moderately tolerant to white rust and *Alternaria* leaf blight, low aphid infestation.
Kesari Gold	90-100	1.79	West Bengal	Suitable for irrigated conditions in *rabi* season, oil content 40.0–42.3%, moderately resistant to white rust; tolerant to aphid infestation.
Kesari 5111	92–105	1.48	West Bengal	Suitable for irrigated conditions in *rabi* season, oil content 41.1–42.5%, moderately resistant to white rust and moderately tolerant to aphid infestation.

Variety	Duration (days)	Yield (t/ha)	Area of adoption	Salient features
Bayer Mustard 5222	105–122	1.68	West Bengal	Suitable for irrigated conditions in *rabi* season, oil content 38.0-39.1%, moderately resistant to white rust and tolerant to aphid infestation.
TBM-204 (Trombay Bidhan Mustard-204)	110	1.34	West Bengal, Bihar, Jharkhand, Odisha, Assam and Manipur	Suitable for irrigated conditions in *rabi*/winter season, oil content 41%, moderately resistant to *Alternaria* leaf spot.
SVJ-64	125–130	2.62	Haryana	Suitable for irrigated conditions under both high and low fertility conditions, oil content 39.1%
Pant Girija (PYS 2012-6)	83–128	1.44	Uttarakhand	Suitable for timely sown irrigated conditions for Uttarakhand plains, oil content 45.30%, resistant to *Alternaria* blight, white rust, downy mildew and *Sclerotinia* rot.
Shalimar Sarson-2 (KBS-49)	205–215	1.5–1.6	Irrigated conditions of Jammu & Kashmir at 1,500–1,800 m above MSL	Oil content 42.7%
Shalimar Sarson-3 (KBS-3)	200–210	1.5–1.7	Irrigated conditions of Jammu & Kashmir at 1,800–2,000 m above MSL	Oil content 41.3%

4.2 Sunflower

Sunflower is mainly grown for its oil. The oil is used for culinary purposes, in preparation of vanaspati and in the manufacture of soaps and cosmetics. Sunflower oil is light in colour, low in saturated fats, has a neutral taste and is ideal for cooking because it can withstand high temperatures. It is especially recommended for heart patients. Its cake is rich in protein and is used as a cattle and poultry feed. Today sunflowers are one of the world's leading oil seed crops, second only to soybeans. Sunflower oil for cooking purposes is marketed worldwide. It is also used as bio-diesel, or a vegetable-oil based fuel used for running many vehicles, including farming equipment. The benefits of bio-diesel are many. It burns 75% cleaner than petroleum based diesel. It is highly lubricating, reducing wear and tear on engine parts. It requires no engine modifications to be used, and can even be mixed with petroleum based fuels. The by-product of bio-diesel is glycerin, which can be used in the manufacture of soap or hundreds of other products. One of the most beneficial uses of sunflowers is in the removal of toxic waste from the environment. Possessing extensive root systems, they are able to reach deep into sources of polluted water and extract large amounts of toxic metals, including uranium.

Uses of Sunflower

1. Edible oil of sunflower is of superior quality because of anti-cholesterol and non- cholesterol contents and is thus preferred over other oils for cooking food.
2. Its oil is used for making vanaspati, soap and other allied products.
3. Its oil is used in the manufacturing margarine.
4. Sunflower meal in the form of cake is used for animal feed or manure.
5. Hulls and stalks are used for the preparation of furfural.
6. Green plants are used as forage or silage.
7. Dry stalks are used as fuel.
8. Pulp is also extract from plants.
9. Sunflower husk finds use in the preparation of ethyl alcohol and yeast.
10. Salted and fried kernels of sunflower are eaten as such and its meal is used for confectionery.
11. The concentrates and isolates of defatted meals are used for protein additives, meat analogues, meat extenders, emulsifying agents, bakery products, beverages, snack products, etc.
12. Seed is included in many commercial bird feeds.

Table 4.3: Nutritional quality of sunflower oil

Qnalities	Percentage
Total digestible nutrients	67.0
Crude protein	11-12
Ether extract	10-12
Crude fiber	31.0
Acid detergent fiber	32.0
Lignin	10-16
IVDDM 2	63-70

Sunflower oil margarine has become a popular alternative to butter. In Spain sunflower seeds are roasted in their shells and enjoyed as a snack. They are rich in vitamins B1, B3, and B6 and are also made into a nutritious spread, available from health food stores. All parts of the sunflower are useful: the leaves make good cattle fodder, while the fibrous stems were once used to make paper. The young flower buds can also be boiled and eaten like artichokes.

Medicinally, sunflower seeds have expectorant and diuretic properties and were once considered beneficial for colds, coughs and bronchitis. The leaves and flowers, which have weak insecticidal properties, were thought to offer some protection against malaria, and a poultice of the fresh leaves was a Russian folk remedy for fevers.

Table 4.4: Constituent of sunflower oil

Sl. No.	Constituent	Percentage
1.	Protein	20.8
2.	Lipid	54.8
3.	Carbohydrates	18.4

4.2.1 Origin and Distribution

The cultivated sunflower (*Helianthus annuus* L.) is a native of southern United States and Mexico, from where it was taken to Spain before the middle of the sixteenth century. Sunflower was considered as an ornamental flower until 19th century after that it was cultivated as an oil seed plant in Russia. The botanical name *Helianthus* means flower (anthos) of the Sun (helios), a reference to the round yellow heads that resemble the sun's disk surrounded by rays. The flower heads are also supposed to rotate so that they always turn towards the sun.

In the nineteenth century, the cultivation of sunflower as an oilseed crop began in the Soviet Union and the majority of the present day varieties grown all-over the world trace back their origin to the USSR. The major sunflower-producing countries of the world are the Soviet Union, Argentina, Bulgaria, Turkey and South America. Sunflower, as an oilseed crop, was introduced into India in 1969, however, commercial cultivation started in the year 1972 with the introduction of open pollinated variety of Russian origin.

4.2.2 Botany

Sunflower belongs to the family Compositae and the genus *Helianthus*. The genus *Helianthus* comprises about 67 species-all native to America; out of these 67, two species, viz. *H. annuus* and *H. tuberosus* are cultivated. The commercial varieties cultivated for seed purposes are grouped under *Helianthus annuus* L.

variety macrocarpus. The common cultivated sunflower, i.e., *H. annuus* is a diploid, contains 34 somatic chromosomes (2n=34). The cultivated genotypes are characterized by a single stem terminating in a capitulum. Sunflower is protandrous, in which the male and female elements mature at different times. There appears to be a time-lag of 18-24 hours in the maturity of the male and female elements. So it is essentially a cross-pollinated plant, besides showing varying degrees of self-incompatibility. The species *H. tuberosus* called Jerusalem artichoke is a hexaploid (2n=102) and cultivated for tubers.

4.2.3 Climatic and Soil Requirement

Sunflower is a day- neutral plant can be grown successfully in different seasons under varying conditions of day length, provided the temperature is favourable. Sunflower, basically a temperature plant, can also be grown commercially in tropical or sub-tropical conditions. It grows best with clear sky and occasionally rain shower during early stages. This crops requires a cool climate during germination and seedling growth and warm non-cloudy weather and high temperature during flowering to maturity. Sunflower can tolerate temperature range of 8 to 30 °C but the temperature below 16 °C and above 40 °C reduces yield and oil content. High temperature affects pollination and fertilization, thus care should be taken to adjust the sowing date in such a way that flowering does not coincide with high temperature. The optimum temperature for best performance under controlled conditions is 27-28 °C. Sunflower can be planted successfully when soil temperature exceeds 100 °C, though germination begins at 3-5 °C. Young plants (1 or 2 pairs of leaves) can withstand frost to -6 °C. However, plants are frost sensitive from 6-8 leaf stage until flowering finishes. Late frost does not destroy the whole plant but destroys the fruiting head, leading to increased branching, resulting in small heads with many empty fruits. During flowering temperature should be in between 18-20 °C. Goods yields are obtained where the ruling temperature is 18-22 °C showing that generally speaking the crop needs a warm climate. Average minimum temperature below 17 °C between flowering and maturity is necessary to ensure best oil quality. The crop is grown from 45° S to 55° N but greater production is between 20-50 °S. The crop can be cultivated up to an altitude of 2500 m but highest yield of oil is obtained below 1500 m. Sunflower cannot tolerate frost and requires a frost free period of 120 days. It is resistant to drought but requires continuous availability of soil moisture for the best performance.

Sunflower can thrive on a wide range of soils such as sandy loam, black soil and alluvial soil, but it performs better on sandy loam soil than on clay soil under similar management. However, good drainage is more important than soil type. Thus, fertile, well drained, medium textured soils having heavy water holding capacity are best suited for the cultivation of sunflower. Sunflower is highly susceptible to water-logging. Avoid cultivation of sunflower on soils affected by salts and with poor drainage. The optimum pH requirement is 6.0 to 8.5.

4.2.4 Variety

Promising varieties namely: MSFH-17, MSFH-1051, MSFH-8, Surya-51, PSFH-67, GKSFH-2002, Jwalamukhi, Mega-363, PAC- 302, PAC- 304, PAC-36, PAC-3776 and NSFH-592 have been recommended. These hybrids are medium tall with dark green leaves; their seeds are black in colour and mature in about 110-125 days. Mega-363 is a long duration hybrid and hence should not be selected for cotton areas. These hybrids have an average yield about 2 tonnes per hectare with oil content of 40.5 to 43.0 percent. In hybrids fresh seed is to be sown every time which adds to its cost of cultivation.

Andhra Pradesh: APSH-11, PAC-36, PAC-1091, Modern

Tamil Nadu: MSFH-8, KBSH-1, Sungene-85, CO-1, CO-2, TCSH-1

Punjab and Haryana: KBSH-1,Jwalamukhi, Sungene-85, PAC-36, LS-1

Gujarat: KBSH-1, Jwalamukhi, Sungene-85, PAC-36, DSH-1

West Bengal: Surya-51, PSFH-67, GKSFH-2002, PAC- 302, PAC-304, PAC-36, PAC-3776, MSFH-17, MSFH-1051, MSFH-8, PAC-361, Aditya, JK Chitra, Sunbred-275

Recently released varieties / hybrids of sunflower

Variety	Year of release	Yield (kg/ ha)	Days to maturity	Oil (%)	Area of adaptability	Salient features
Modern	1978	800-1500	80-85	34-35	All India	Dwarf and early variety
SS-56	1988	800-1400	82-88	34-36	Maharashtra	Early variety
GAUSUF 15 (Guj. Sun-1)	1993	1200-1800	92-98	35-39	Gujarat	Tolerant to *Alternaria* leaf spot and rust
TNAU-SUF-7	1995	1000-1700	85-90	38-41	All India	-
PKVSF-9 (AKSF-9)	1996	1300-1600	82-85	38-40	Vidharba region of Maharashtra	Moderately resistant to downy mildew and *Alternaria*
LS-11 (Sidheswar)	1998	1000-1800	85-90	36-39	Maharashtra	Tolerant to downy mildew, rust and *Alternaria*
DRSF-108	2004	900-1800	95-100	36-39	All India	High oil content
CO-5	2005	1000-1700	85-90	39-42	Tamil Nadu	-
TAS-82	2005	800-1200	90-100	40-42	Maharashtra	Black seeded variety
LSF-8	2006	100-1400	90-95	36-39	Maharastra	Tolerant to downy mildew, rust and *Alternaria*
DRSF-113	2007	1000-1500	90-98	36-39	All India	High yield
Phule Raviraj	2009	1795	90-95	34	Western Maharashtra	Tolerant to necrosis, *Alternaria* and capitulum borer
BSH-1	1980	900	85	41	All India	Resistant to rust and downy mildew
APSH-11	1987	1200	90-95	41	AP	-
LSH-3	1988	1200	95	39	Maharashtra	Resistant to downy mildew
KBSH-1	1992	1400	90	43	All India	High oil content
PKVSH27	1993	1300	95	41	Punjab	-
PSFH-67	1996	1300-1400	85-90	39	Vidharba region of Maharashtra	Moderately resistant to downy mildew
DSH-1	1997	1300	85	40	North Karnataka	Resistant to downy mildew

Variety	Year of release	Yield (kg/ha)	Days to maturity	Oil (%)	Area of adaptability	Salient features
TCSH-1	2000	1500	85-90	39	Tamil Nadu	Tolerant to *Alternaria* and rust
KBSH-41	2002	1300-1500	90-92	39-41	Karnataka	Tolerant to moisture stress
KBSH-42	2002	1300-1500	90-92	38-41	Karnataka	Tolerant to moisture stress
PSFH-118	2002	1400	85-88	40	Punjab	Resistant to stem and head rot
KBSH-44	2003	1400-1600	95-98	36-38	All India	Resistant to downy mildew
LSFH-35 (Maruti)	2003	1400-1500	90-95	39-41	Maharashtra	Resistant to downy mildew
NDSH-1	2003	1400	88-90	40	Andhra Pradesh	Early maturing hybrid
RSFH-1	2004	1300-1500	95-100	39-41	Karnataka	High oleic hybrid
HSFH-848	2005	1800-2400	90-95	41-42	Haryana	Dwarf hybrid
DRSH-1	2006	1300-1600	92-98	42-44	All India	High oil content
KBSH-53	2009	1700-2700	95-100	42-44	Karnataka	Resistant to powdery mildew
PSFH-569	2009	2232	95-100	40	Punjab	High oil content and early hybrid

4.2.5 Agronomic Management

4.2.5.1 Land preparation

Sunflower requires a fine seedbed, free of weeds and clods, for proper germination and good crop stand. To bury the residues of previous crop, ploughing with disc harrow followed by 2-3 cultivations and plankings may be required. Fields vacated by potato may not require much seedbed preparation. However, adequate soil moisture is essential for getting good seed germination.

4.2.5.2 Sowing parameters

Spring season is the best for getting higher and stable seed yield sunflower. Sunflower being a cross pollinated crop depends on honey bees for pollination and seed setting. Availability of honey bees is abundant during spring season due to favourable weather condition. Planting during spring should be made after the soil temperature exceeds 10°C at 10 cm depth at 8 a.m. Early planting at lower temperature increases time of flowering without much earliness. Best results are obtained when crop is sown in the first fortnight of January which produces higher seed yield than later planting. Sowing of short duration hybrids, however, can be extended up to second fortnight of January. Besides low yields, the late sowing may also face early monsoon rain at harvesting and threshing time causing difficulty in its management.

Use 5 kg seed per hectare for hybrids to achieve the required plant stand. Low seed rate recommendation for hybrids is mainly because of smaller seed size and wider spacing for hybrids. Seed should be treated with Mancozeb or Carbendazim @ 2 g per kg of seed to control seed and soil borne fungal diseases.

A spacing of 60 cm x 30 cm for hybrids has been found to be an optimum. The spacing result in better development of head, more seed filling and less stem breakage.

Crust Breaking

Rain after sowing results in crust formation, which hinders the proper emergence of seedlings. Thus it may be necessary to break the crust, which can be done with harrowing. Harrow as soon as possible after the crust has performed and if possible, before seedlings have emerged.

4.2.5.3 Nutrient management

Fields studies have shown that hybrid sunflower responds well to nitrogen application up to 60 kg/ha, further increase in nitrogen results in lower seed and oil yield. Application of P fertilizers @ 30 kg P_2O_5/ha increase sunflower yield. Sunflower responds to K application only on soils deficient in available potassium and it is recommended to be applied @ 30 kg K_2O/ha on such soils.

Regarding application method, drilling the fertilizers besides the seed row has been found to be more beneficial for getting higher seed yield. Application of fertilizers below the seed was found to be injurious to germinating seeds and reduced seed emergence and seed yield.

Entire dose of N, P and K should be applied at the same time of sowing. Prefer single super phosphate for P application as it also contains sulphur to which sunflower responds favourably. If single super phosphate is not available then apply gypsum @ 25-30 kg/ha at sowing time. A supplementary dose of nitrogen @ 25-30 kg/ha may be top-dressed a month after sowing sunflower if the crop exhibits symptoms of nitrogen deficiency. Nitrogen deficiency appears as yellowing of the older leaves, which die prematurely. Leaves are also fewer and smaller. Plants with a severe deficiency are yellow and stunted, while young plants are of purplish colour. Plants are slender with longer internodes. The flower head is small with many unfilled seeds and the oil content is reduced.

Phosphorus deficiency results in delayed flowering, stunted growth and small leaves which prematurely die back from the leaf tip. Phosphorus deficiency is shown in small dark green leaves later turning pale with necrosis, growth is restricted with short stiff stems, flowers are small, stunted malformed, ripening is delayed; oil content and yield are reduced.

Potassium deficient plants are squat with a proportionately greater reduction in stem length then in leaf size. Initially leaves are dark green but turn yellow at the edges and die except the lower leaves. Potassium deficient plants have restricted growth with a rosette arrangement of the younger leaves; later the leaves develop dark brown necrotic patches.

4.2.5.4 Water management

Sunflower can withstand moderate drought because of its deep root system. However, it responds favourably to irrigation and its seed yield can be increased substantially by sustaining adequate moisture in soil by timely irrigations besides improvements in oil content. Sunflower is highly moisture sensitive plant and its broad leaves exhibit wilting with slight moisture deficiency.

First irrigation to sunflower should be applied a month after sowing and interval during the subsequent months be reduced to 2-3 weeks during March and 8-10 days during the hot months of April and May. Irrigation should be stopped about 12-14 days before harvest. Sunflower is a tall growing plant with heavy head, therefore, irrigation at flowering and thereafter should be given on calm days to avoid lodging. The incidence of *Sclerotinia* stem-rot is more under irrigated conditions especially where the soil drainage is poor. Excessive and frequent irrigations on fine textured soils should be avoided.

4.2.5.5 Weed management

Mechanical methods such as hoeing at 3-4 weeks after sowing followed by earthing up before flowering has been found to be effective in checking weeds. Use of tractor drawn implements before the crop attains 60-70 cm height has also been found effective. For better efficiency of manual weeding an improved 3 tine wheel hand-hoe could also be used. Removal of weeds through mechanical/manual means are quite effective but expensive, moreover, often timely labour is not available. The studies initiated to find effective herbicides for weed control in sunflower, revealed that pendimethalin (STOMP @ 2.5 litres/ha) as pre-emergence application gave effective weed kill.

4.2.5.6 Isolation distance

The distance maintained for successful production of seed should be 400 m from field of other varieties and fields of the same variety not conforming to varietal purity requirements for certification under foundation seed, and 200 m from field of other varieties and fields of the same variety not conforming to varietal purity requirements for certification under certified seed.

4.2.6 Allelopathic effects of Sunflower

Researches show that sunflower depresses the growth and productivity of succeeding crops. The reason postulated are the allelochemicals being secreted by sunflower crop. Decomposing residues of sunflower could also bring about allelopathic effects. Phytotoxic compounds released into the soil rhizosphere have been observed to cause varying degree of allelopathy/ autotoxicity in crop. Field studies conducted in the Department of Agronomy, Punjab Agricultural University, Ludhiana on the allelopathic effect of spring sunflower on succeeding *Kharif* crops revealed that maize, sorghum, cowpea, pearl millet, pigeon pea, mung bean and sesamum suffered 41.7, 36.7, 27.6, 28.8, 23.0, 10.8 and 9 percent reduction in yield, respectively, as compared to these when grown after wheat. Studies in allelopathic effects of different plant parts of sunflower indicated that sunflower roots reduced pigeon pea, cotton and mungbean yields the most, followed by roots plus shoots treatment. Sunflower shoots did not have any harmful effect on growth of these crops. Bioassay studies involving use of aqueous sunflower extract, decomposing plant residue and soil from area in proximity of sunflower crop revealed inhibition of germination and growth of cotton, pigeon pea, sunflower and mung bean. These studies further showed that effect of leaf extract/residue was more than roots. The allelopathic effect of sunflower plant residue persisted up to 24 days after incorporation and beyond this period the effect was rather meagre. Chromatograhic and physico-chemical analysis of allelochemicals in sunflower parts showed the presence of chlorogenic and isochlorogenic acids in both stems and leaves besides these some unidentified phytotoxins were also observed. The concentration of these was more in leaves than in stems. Further studies revealed that total phenolics were more in leaves followed by in roots and shoots. Sunflower raised under moisture stress conditions had higher phenolics as compared to the raised without any moisture stress.

4.2.7 Insect Management

Cut worm (*Agrotis* spp.): It is a polyphagous pest and cuts off the tender stalks, near ground, of young crop plants during night. During day time the caterpillars remain hidden in the soil. It is more serious during March-April when sunflower follows potato. When flat sowing is practieed, apply 5 liters of Dursban/Radar 20 EC (Chlorpyriphos) per hectare before sowing. These insecticides should be mixed in 25 kg fine soil and broadcast uniformly after last ploughing but before planking. The crop is to sown on ridges to avoid cut worm damage in the early stages.

Tobacco caterpillar (*Spodoptera litura*): The polyphagus pest remain active throughout the year feeding on one or the other crop. The newly hatched caterpillar is pale-greenish with a black head and feeds on the outer epidermis of leaves. It poses a serous threat during April-May and defoliates the plants.

Cabbage Semi-looper (*Plusia orichalcea*): This insects feeds on many crops including the sunflower. The newly hatched caterpillars defoliate the crop and cause severe damage to it during April-May.

Bihar hairy caterpillar (*Spilosoma obliqua*): It is also called Bhabbu Kuta. This is highly polyphagous and voracious feeder. Newly hatched larva is bright brown and its body is covered with hair and legs are brown. At the older stage it is much darker and with dark brown hair of almost uniform length. these insects feed gregariously and should be collected along with the plant and buried deep in soil.

For the control of above mentioned pests, spray 1.125 L Folithion/Sumithion 50 EC (fenotrothin) or 1.225 L Thiodan 35 EC (endosulfan) or 500 ml Nuvan 100 EC (dichlorvos) in 250 litres of water per hectare. Spraying of insecticide is to be repeated after two weeks, if necessary.

Head borer (*Helicoverpa armigera*): Head borer or American boll worm has become the noxious polyphagous pest in India and also causes heavy losses in sunflower. The full grown caterpillar is brownish green having grey longitudinal streaks on it. The adults lay eggs on the flower buds. After hatching the larvae feed on developing seeds.

Head borer can be controlled by spraying 2.5 kg of Sevin 50 WP (Carbaryl) or 2.0 litres of Asataf 75 SP (Acephate) or 2.125 litre Folithion 50 EC (Fenitrothion) or 2.5 litres of Dursban 20 EC (Chlorpyriphos) or Thiodan 35 EC (Endosulfan) or 1.25 litre of Nuvacron 30 SL (Monocrotophos) in 375 litres of water per hectare. Spraying of insecticide is to be repeated after two weeks, if necessary.

Endosulfan is comparatively safe for honey bees. Spray when honey bee activity is minimum i.e., early morning or late evening.

4.2.8 Pollinators and Birds

Bees are beneficial to sunflower yield because they carry pollen from plant to plant which results in cross pollination. Some sunflower varieties will not produce highest yields unless pollinators are present. All varieties will produce some sterile seed (without meats), but varieties differ in their degree of dependence on insect pollinators. Autogamous sunflower hybrids do not require bees for maximum yield and will yield the same when covered by bags as uncovered. In non-autogamous sunflower varieties, pericarp (bull) development is normal but no ovules or meat developed. Wind is relatively unimportant in cross-pollination of sunflower. Some of the older open-pollinated varieties set only 15 to 20% of seed without pollinators, whereas many hybrids set 85 to 100% seeds without pollinators.

Insect pests have become major potential yield-reducing factors in sunflower production in the northern Midwest. Insects specific to sunflower that feed on the heads include the larvae of three moths; sunflower moth, banded sunflower moth and sunflower bud moth. Sunflower midge has caused widespread damage in some years. Sunflower head clipping weevil, sunflower beetle, sunflower maggot, wireworm, grasshopper, cutworm, sugarbeet webworm, ragweed plant bug, wooly bear and painted lady caterpillar have caused occasional damage to sunflower. Adults of insect pests of other crops (such as corn rootworm beetle and blister beetle) can be found as pollen feeders on sunflower heads, but usually cause little injury.

Precautions to be taken for spraying insecticides

1. Spraying of insecticide during flowering should be avoided to protect the pollinators.

2. Spraying should be done in the afternoon when pollinators are less so that there is a time gap till morning or mid day when their activity is more.
3. Spraying should be done only when pest incidence is observed in abundance.
4. Spraying should be done on calm day to avoid drift.
5. Precautionary measures are to be taken if pest outbreak is observed every year.

4.2.9 Disease Management

Steam rot / cottony rot: It is soil and seed borne disease. Infestation of basal portion of stem takes place via mycelium that grows with the roots and the stem becomes soft and cotyledon leaves flatten on the ground. At bud formation soft decaying area appears on its stem and base and white mycelia appear later on. The stem becomes weak, leaves dry up and drop-off with strong wind.

Infestation takes place on heads when fully developed and seed formation has taken place. The back of head becomes soft and turns brown. Fungus like cotton appears and seeds are also infested. The sclerotia of this pathogen, almost of the size of sunflower seed are produced in abundance which are visible in pith of rotten stem and also in the infested head. The affected tissue turns black and shredded. This disease can be effectively controlled by treating seed with Thiram @ 2 g/kg seed before sowing.

Root rot or charcoal rot: It is soil borne disease that infects the roots. Roots turn brown and rot. Infection at early stages kills the seedlings rapidly. This also causes neck rot i.e. stem rotting immediately behind the head where infected stem turns brownish and darker towards the base. Roots turn grey and stunted. At later stages, it causes premature ripening, smaller and poor filled heads and a lower yield. For its control, treat the seed with Thiram @ 2 g / Kg seed and avoid moisture stress.

Head rot: High humidity and injury to plants favour infection at lower surface of head where it is attached to the stem. Irregular water soaked spots appear, which gradually enlarge and become soft. Shredding of back of heads takes place. Cracking of front side is also common. Seed filling is affected severely and seeds remain small and shrivelled. The yield losses range from 20 to 60 percent and oil reduction to the tune of 25-53 percent and increase in free fatty acids, which is undesirable. To avoid this disease plough the soil with furrow turning plough to bury the debris of previous crop and residues of sunflower; treat the seed with Thiram @ 2g/kg seed; avoid excessive stress of irrigation;

grow recommended hybrid and don't grow sunflower repeatedly in the same field; Spray with Indofil M-45 @ 625 g per hectare in 250 litre of water at 15 days interval starting at about 60 days stage.

4.2.10 Harvesting and Threshing

The crop is ready for harvesting when the backs of flower heads turn lemon yellow or yellowish brown near the stalk and the discs start drying. At this stage the seeds give a blackish look and are fully ripe. The shedding of disc-flower residues from undamaged heads also indicate physiological maturity. Promptly harvesting seed after physiological maturity has many advantages ; seed is least exposed to bird and rodent attack; the crop cycle is shorter resulting in more time to prepare for the next crop ; reduces harvest seed losses ; provides high quality seed and reduces exposures to bad weather. The harvested heads are mildly dried and threshed by beating or trampling with tractors or sunflower threshers. The sunflower threshers are more efficient and economical. After threshing the produce should be thoroughly dried before storing otherwise in storage it gets affected by fungus and decomposes.

4.2.11 Yield

Rainfed crop : 0.8-1.2 t/ha and Irrigated crop : 1.8-2.3 t/ha

4.2.12 Post-Harvest Handling

Drying of seeds is the first step in sunflower seed processing. The moisture content of freshly harvested sunflower seeds may be as high as 20%. To ensure safe storage, the seeds must be dried to less than 10% moisture. Sunflower seed is decorticated before the oil is extracted to ensure that the cake is a nutritious animal feed.

Extraction of oil from sunflower seeds or kernels can be done using general equipments and operating conditions used for soybean or other oil seeds. The extraction of oil from sunflower is done by mechanical extraction, prepress solvent extraction and direct solvent extraction methods. Sunflower oil usually does not require extensive refining as it contains relatively low levels of free fatty acids, phopholipids, tocopherols, pigments and sterols.

Flow Chart for Sunflower Oil Extraction

Though regarding seed production point of view there is no need of oil extraction but as a general purpose it is briefly discussed.

Sunflower seeds → Sun drying → Cleaning by cleaners (by using a set of sieves) → Aspirator (to remove light particles or other unwanted materials) →

Dehuller or Craker (to crack the seeds or to remove the outer cover) → Sieving → Rolling or Exfelling

↙ ↘

Oil Oil Cake

↓

Refining

Important Points to be Considered for Obtaining Higher Yield

- Continuous cropping of sunflower year after year in the same field should be avoided.
- Avoid low-lying areas.
- Always use certified seed obtained from authorized seed producing/ supplying agencies.
- Always treat the seeds with recommended pesticides before sowing.
- Always maintain optimum plant stand recommended for particular region. Maintain one seedling per hill by thinning at 15-18 DAS.
- Incorporate FYM or Compost 2-3 weeks before sowing.
- Spray boron @ 0.02% at ray floret opening stage.
- Protect the predators such as Coccinellids and Chrysoperla.

To increase the coverage under sunflower the following steps are necessary.

- Encourage only high yielding sunflower hybrids.
- Make available quality, disease-free, processed, treated seed.
- Best extension services to educate farmers and to provide package of practices.
- Price support mechanism must be in operation and be effective all time.

We Need to Avoid

- Promoting sunflower round the year
- Crop after crop of sunflower
- Promoting sunflower in any type of soil
- Promoting low yielding varieties.

Advantage of Taking Sunflower in High Intensity Cropping System

- Being a short duration crop, it can be well suited as catch crop.
- It is a drought, frost and salt tolerant crop.
- There is very slow degeneration in seed quality hence the same seed can be used for sowing up to 4-5 years.
- Sunflower is a useful crop for apiculture and it attracts the bees and provides them nectar.

Problems In Growing Sunflower

- Greater menace of the birds specially of crows and parrots.
- Poor filling of seeds specially of disc florets.
- Unavailability of ideo-types and their superior seeds.
- Poor viability and lesser keeping quality of seeds.
- Poor market and acceptability of raw and fresh oil by customers.
- Oil becomes bitter if not extracted within 90 days of harvesting.

Potential Uses for Sunflower Seed Products

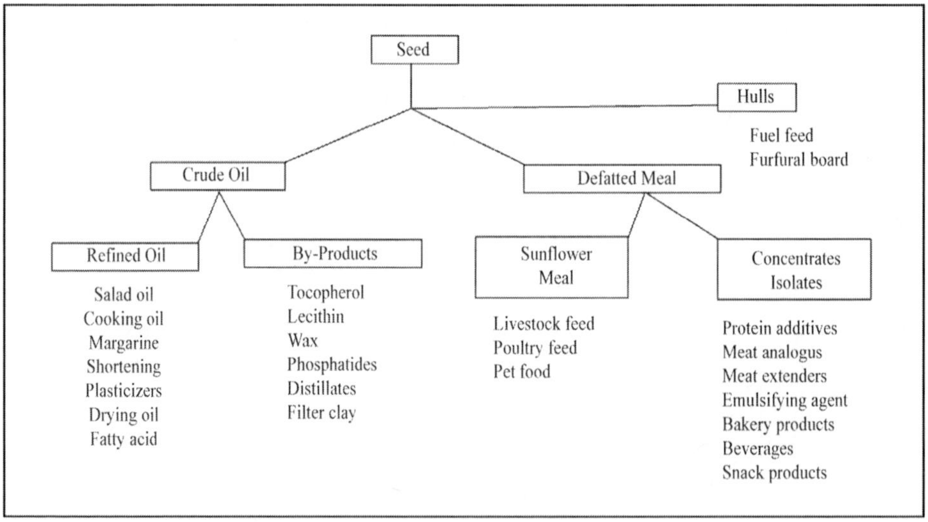

4.3 Groundnut

Out of the nine oilseeds crops grown in our country, Groundnut is the only crop cultivated in Goa during both the seasons. The major area under this remunerative crop (1437 ha) however is during the *Rabi* season. Although this crop has limited acreage at present, there is good scope for bringing more area under this crop since the per unit productivity ranks third in the country (1.9 t/ha) as against the National average productivity of (1.0 t/ha).

The oil content of the seed varies from 44 to 50 per cent, depending on the varieties and agronomic conditions. Groundnut oil is an edible oil. It finds extensive use as a cooking medium both as refined oil and Vanaspati Ghee. It is also used in soap making, and manufactoring cosmectics and lubricants, olein stearin and their salts. Kernels are also eaten raw, roasted or sweetened. They are rich in protein and vitamins A, B and some members of B_2 group. Their calorific value is 349 per 100 g. The residual oilcake contains 7 to 8 per cent of N, 1.5 per cent of P_2O_5 and 1.2 per cent of K_2O and is used as a fertilizer. It is an important protein suppliment in cattle and poultry rations. It is also consumed as confectionary product. The cake can be used for manufacturing artificial fibre. The haulms (plant stalks) are fed. Groundnut is also of value as rotation crop. Being a legume with root nodules, it can synthesize atmospheric nitrogen and therefore improve soil fertility.

Why Grow Groundnut?

Cultivation of any crop must accrue maximum benefits to the grower not only in terms of the returns but also other indirect benefits associated with that crop. Groundnut being a leguminous crop, has the following advantages over other crops including the horticultural crops.

1) Being leguminous, groundnut has the ability to fix atmospheric nitrogen biologically into the soil which enriches the soil and this benefits the succeeding crop.
2) It is one crop which does not require irrigation and can be cultivated on the residual soil moisture conditions.
3) It does not need elaborate management practices.
4) It can provide gainful employment during the lean season.
5) Unlike the vegetables and fruits, which are highly perishable items, groundnuts can be stored for a long period of time and sold when prices are high.
6) It provides a very nutritive biomass in form of green fodder which can be fed to cattle.
7) Groundnut crop is considered not only as an oilseed crop but also as a food crop because of varieties containing high protein are available.

4.3.1 Origin

Groundnut (*Arachis hypogaea* L.) is believed to be the native of Brazil to Peru, Argentina and Ghana, from where it was introduced into Jamaica, Cuba and other West Indies Islands. The plant was introduced by Portuguese into Africa from where it was introduced into North America.

4.3.2 Botany

Groundnut (*Arachis hypogaea* L.) is a member of sub-family, Papilionaceae of the family Leguminosae. *Arachis hypogaea* L. consists of two subspecies each containing two botanical varieties. Subspecies *hypogaea* variety hypogaea (The Virginia Group) variety *hirsuta* Kohlar Subspecies *fastigiata* Waldron variety *fastigiata* (the Valencia Group) variety *vulgaris* (the Spanish Group)

Plants of the botanical variety *hypogaea* are spreading (runner) to upright (erect bunch) in growth habit, have alternate branching, lack inflorescences on the main stem, possess appreciable fresh seed dormancy, flowers are longer and mature later than those of subspecies fastigiata. Variety *hirsuta* has been used only to a little extent.

Plants of this subspecies fastigiata are upright, have sequential branching and inflorescences in the main-stem leaf axils, possess little fresh seed dormancy, and are of shorter duration than those of the subspecies *hypogaea*. Subspecies *fastigiata* includes both the Spanish and Valencia types

Groundnut, in general, has a short-statured plant, with the main axis being upright (15 to 40 cm long) but the major part of the plant consists of the primary branches. Secondary and tertiary branches are found in the semi-spreading and spreading (Virginia) types, giving them a prostrate stature. The leaves are alternate, stipulate and quadri-foliate. The flowers are orange yellow, typically papilionaceous, with a long calyx tube within which is held the style borne on a superior ovary. The calyx and corolla lobes are borne in the axils of the leaves on the fruiting branches. They are bisexual, zygomorphic, complete and sessile. Petals are five, with one large standard, two wings and two fused keel petals. There are two steering and eight fertile anthers, four of which are globose and four oblong (dimorphic).

Groundnut is predominantly a self-pollinated crop and pollination takes place early in the morning. As soon as the fertilization is complete, the flowers fade. After fertilization, an intercalary meristem becomes active at the base of the ovary above the point of attachment of the hypanthium, producing new tissue below itself and resulting in an elongated stalk in the peg. The pegs are positively geotropic, enter the soil and bend in a horizontal plane. Generally two, and occasionally one, three or four, fertilized ovules are borne at the tip of the peg which later swells to become the pod. The testa is generally pink, but varieties with red, white, purple and blotched testa, with various gradations of colours are available.

4.3.3 Variety

There are three types of varieties in groundnut, bunch types with erect plant habit, spreading and semi-spreading types. The bunch types have light green foliage, comparatively broad leaflets and mature early. However, they are usually susceptible to tikka disease. The spreading types usually have dark green foliage with smaller leaflets. These are usually late in maturity. The semi-spreading varieties are intermediate between the bunch and the spreading types. A number of promising varieties have been evolved in various groundnut growing states for commercial cultivation.

Description of some of the important varieties of groundnut is given below:

Jyoti: It is a bunch type of variety with dark green foliage. It matures in 95-110 days. It is susceptible to tikka disease. It is suitable for growing in Madhya Pradesh. It yields about 12-15 quintals per hectare under rainfed conditions and 20-22 quintals per hectare under irrigated conditions. Seeds are bold and contain 53 per cent oil. Its shelling out turn is 78 per cent.

RS-1: It is a spreading variety. It matures in 135-140 days. It is tolerant to tikka disease. It is suitable for growing on sandy soils of Rajasthan. It yields about 15-20 quintals per hectare. It has 77 per cent shelling out turn. Seeds are of medium size and contain 48 per cent oil.

Type-28: It is a spreading variety. Foliage is of dark green colour. It matures in 125-135 days. It is tolerant to tikka disease to some extent. Seeds are of medium size. Its yield potential is 20-25 quintals per hectare. Its shelling out turn is 72 per cent. It contains about 48 per cent oil. This variety is suitable for growing in Uttar Pradesh.

Type-64: It is a semi-spreading variety. Foliage is of dark green colour. It matures in 115-120 days. It is also tolerant to tikka disease to some extent. Seeds are bold and attractive. Its yield potential is 20-25 quintals per hectare. Its shelling out turn is 70 per cent. It contains about 49 per cent oil. It is suitable for growing in Uttar Pradesh.

Chandra: It is a spreading variety. Foliage is of dark green colour. It matures in 125-135 days. Its yield potential is 20-30 quintals per hectare. Seeds are bold and contain 48 per cent oil. Its shelling out turn is 72 per cent. It is very suitable for growing in Uttar Pradesh.

Chitra: It is a semi-spreading variety. Foliage is of dark green colour, It matures in 125-130 days. Its yield potential is 25-30 quintals per hectare. Seeds are of medium size and contain 49 per cent oil. Its shelling percentage is 72, It is suitable for growing in Uttar Pradesh. Groundnut production is highly vulnerable to rainfall deviations and display huge fluctuation between years.

Gujarat (1-3.5 million tonnes), Tamil Nadu (1million tonnes), Andhra Pradesh (1-2 million tonnes), Karnataka (0.5 million tonnes), Maharashtra (0.5 million tonnes) are the major producers of seed. Around 75% of the crop is produced in *kharif* (June - September) and remaining 25% in *rabi* (November - March).

According to the Department of Agricultural Research and Education, ICAR, Government of India, (1999-2000) Groundnut CSMU 884 and HNG 10 were released at the national level and LGN 2 and ALR 3 were state release but notified by the Central Variety Release Committee. DRG 12, DRG 17 and ICGS 44 and the breeding lines TSP 60, GB 36 and PBS 12006 can be the source of resistance against bruchid beetle, an emerging serious storage pest in groundnut. In an effort to induce functional male sterility as a means to enhance natural cross-pollination some chemical gametocides were tried. Indole acetic acid among them caused the highest pollen sterility (30.2%).

Table 4.4: Groundnut Varieties Released during 1999-2000

Variety	Plant type/ special features	Maturity duration (days)	Area of recommendation
CSMG 884	Semi-spreading, resistant to early and late leaf-spot and bud necrosis disease	115-120	Uttar Pradesh, Rajasthan, Maharashtra
HNG 10	Semi-spreading, medium bold kernel	115-120	Uttar Pradesh Rajasthan, parts of Madhya Pradesh, Tamil Nadu
LGB 2	Semi-spreading, bunch type	115-120	Maharashtra
ALR 3	Spanish-bunch type, resistant to foliar disease	110-115	Tamil Nadu

Table 4.5: Promising Variety of Groundnut Suitable for Cultivation in Different States.

State	Recommended varieties
Andhra Pradesh	Kadiri-2, Kadiri-3
Bihar	BG-1, BG-2, Kuber
Gujarat	GAUG-1, GAUG-10, PG-1
Haryana	MH-1, MH-2, MH-4, M-145, M-13. PG-1
Punjab	PG-1, M-145, C-501, M-13, M-37, M-197, M-335, SG-84, M-522
Rajasthan	RS-1, RSB-103-87
Tamil Nadu	TMV-8, TMV-9. TMV-10. M-13, Kaushal
Uttar Pradesh	T-28, T-64, Chandra, Chitra, Kaushal, Parkash, Amber
Madhya Pradesh	Jyoti, JL-24
Maharashtra	A.K-12-24, SV-11. Vikram (TG-1), Phule Pragati (JL-24) TG-17, UF70-130, Karad 4-11. Kopergaon-1, Kopergaon-3

Recently released varieties / hybrids of groundnut

Variety	Duration (days)	Yield (t/ha)	Area of adoption	Salient features
Phule Unnati (RHRG-6083)	125–130	3.99 (Summer); 2.85 (*Kharif*)	Maharashtra	Suitable for irrigated and rainfed conditions during summer and *kharif* season, oil content 51%, resistant to *Spodoptera* and rust, tolerant to late leaf spot, and stem rot.
Phule Dhani (JL-1085)	109	3.33 (pod); 2.31 (kernel)	Andhra Pradesh, Telangana, Karnataka and Tamil Nadu	Suitable for cultivation under rainfed conditions during *kharif* season, oil content 50%, highly resistant to foliar diseases (LLS and Rust) and for the combined infestation of *Spodoptera* and *Helicoverpa*.
Gujarat Groundnut 34	111–125	3.72	Gujarat	Suitable for irrigated, agro-climatic variables/ stresses for summer cultivation, oil content 52.8%, tolerant to Tikka and rust diseases, and lower incidence of thrips and jassids.
Dheeraj (TCGS 1073)	105–110 (*Kharif*); 110–115 (*Rabi*)	1.91 (*Kharif*); 2.84 (*Rabi*)	Andhra Pradesh	Suitable for *kharif/rabi* irrigated, timely sown conditions, oil content 48–50%, less susceptible to LLS, rust, PBND and PSND; less susceptible to leaf miner and jassids and other sucking insects.
BSG 0912 (INS I 2013-8) Groundnut BSR 2	105–110	2.22 (*Kharif*); 2.36 (*Rabi*)	Tamil Nadu	Suitable for irrigated tracts, oil content 45%, moderately resistant to rust, late leaf spot diseases, aphids, thrips, jassids and defoliators.
Central-Pragati (TCGS 894)	100-105	2.82 (pod); 1.97 (kernel)	Tamil Nadu, Telangana and Andhra Pradesh	Suitable for timely sown irrigated conditions during *rabi* season, oil content 48%.
Dh 256	113	3.26 (pod); 2.18 (kernel)	Tamil Nadu, Andhra Pradesh, Karnataka and Telangana	Suitable for rainfed conditions during *kharif* season, oil content 50%, tolerant to mid-season drought stress.

Studies were conducted to understand compatibility relationships between the architecture and behaviour of the different habit groups of groundnut and inter-crops. The best genotypic combination was M 13 Virginia type of groundnut + BDN 2 of pigeon pea (tall-medium) which gave the highest total groundnut equivalent yield of 3.63 tonnes/ha. Relative water content and specific leaf area can be taken as important selection criterion for groundnut for identifying high water-use efficiency in drought areas.

4.3.4 Agronomic Management

4.3.4.1 Land preparation

Groundnut crop is mostly grown on the residual moisture conditions. Therefore it is necessary that the soil moisture present in the soil is not exposed to hot sun. However, the land should be cross ploughed with country plough instead of a tractor. Break the clods and remove paddy stubbles. The depth of ploughing must be about 20-30cm.

4.3.4.2 Sowing parameters

The time of sowing assumes great importance in case of *rabi* groundnuts since the crop is taken on residual moisture from the soil. The most ideal time under Goa conditions is between November last week to first week of January. If sowing is delayed beyond this, there is drastic effect on the germination as well as overall yield.

For ensuring optimum plant population, sowing is done by seed drills. However, for sowing with country plough, care should be taken to see that the treated seeds are sown at a spacing of 25cm between rows and 10cm between plants in straight lines. The ideal sowing would be one in which the crop forms a carpet like canopy without any vacant spots. If gaps are seen they could be filled by fresh sowing within ten days time. Optimum plant population in the field is a key to high yields. Seed treatment also plays a vital role in ensuring uniform germination and crop stand.

Seed treatment

Untreated seeds if used for sowing leads to poor germination due to the incidence of seed borne diseases like collar rot, seedling rot, root rot which results in poor plant population and low yields.

Treat the seeds with fungicides like 'Captan' or 'Thiram' @ 3g /kg of seed. For treating the seeds required for sowing one hectare, use 450g of fungicide or the seeds can also be treated with biological fungicide "*Trichoderma viridae*" (Ecofit) in following manner.

For dry seed use 4 g powder/ kg seed. For slurry seed treatment, take small quantity of water @ 10 ml / kg seed and mix 4 g Ecofit and give thorough coating on seeds and air dry before sowing.

For soil application, mix 1 kg Ecofit with 25 kg farm yard manure and broadcast over one acre area. Optimum soil moisture at the time of application ensures best results.

One day before sowing the seed, treat the seed with *Bradyrhizobium* culture @ 300 g /40 kg seed to sow one acre. For better results use rice glue as a carrier.

4.3.4.3 Isolation distance

At least 3 m from irrespective of varieties and types of seed *viz.* foundation or certified.

4.3.4.4 Nutrient management

A starter dose of 10 kg N/ha is necessary for rainfed groundnut at the early stages of crop growth. A top-dressing of additional dose of nitrogen in the form of ammonium sulphate is recommended after 30 days of sowing. It depends upon the growth and effectiveness of root nodules. Application of phosphatic fertilizer to groundnut promotes root growth, multiplication of rhizobium and also helps the crop to tide over the moisture stress. Single super-phosphate is the best source of phosphorus as it contains phosphorus (15%), calcium (17.5%) and sulphur (12.5%). Unless the soil has less than 150 kg K_2O/ha, application of potassic fertilizer is not recommended. The NPK fertilizer doses recommended for the major groundnut-growing states in India are: 15 kg N, 40-60 kg P_2O_5 and 0-45 kg K_2O/ha. All the P_2O_5 and K_2O and half of the N should be incorporated in the soil before sowing. Gypsum is the cheapest source of calcium (25%) and sulphur (18.6%). Well powdered gypsum should be applied on the soil surface as close to the base of the plant as possible when it is in the peak flowering stage. Since, there is little residual effect of gypsum, it is necessary to repeat the application every season. In India, application of gypsum is recommended in the sandy-loam soils @ 500 kg/ha in 2 split doses, half at planting and the other half at the peak flowering stage. Micro nutrients play an important role in stepping up the productivity of groundnut. Application of zinc sulphate to the soil @ 25 kg/ha, once in 3 years has been recommended. When zinc deficiency is noticed in the standing crop, as characterized by yellowing from mid-ribs of young leaves and red spots in the old leaves, foliar application of 0.2% zinc sulphate is recommended. Spraying of 0.1 ppm boric acid (300 mg boric acid in 500 litres water) at 30 and 50 days after sowing resulted in 10 and 15% increased yield respectively. Iron chlorosis in groundnut, prevalent in the calcareous black soils. It is controlled by spraying 1% ferrous sulphate plus 0.1% ammonium citrate.

4.3.4.5 Intercultural operation

Give two hoeings when the crop is about 15-20 days old. In areas where it is possible to do earthing up operation, apply gypsum before earthing up. This should be done before the initiation of flowering which takes place after 45-50 days of sowing. Do not disturb the plants after pegging starts.

4.3.4.6 Water management

The crop is mostly grown on the residual moisture without any supplementary irrigation. However, it responds to two or three irrigations given at an interval of 12-15 days starting from 75 days of sowing, The results from trials show an increase of 35-40% in pod yield corresponding to extra pod yield of 300-400 kg/ha. If the crop is grown in summer (January sowing) irrigations at an interval of 12-15 days would be sufficient.

4.3.4.7 Weed management

Weeds cause maximum damage (30-50% yield loss) to the groundnut crop up to 45 days after sowing. It is advisable to do one or two hand hoeing and weeding depending upon soil type and extent of weed infestation. The earthing up should also be taken up simultaneously with intercultural operations. Basic idea of earthing up is to promote easy penetration of pegs in soil and also to provide more area to spread. Application of herbicides accompanied by 1 or 2 hoeing was found to control the weeds effectively. The recommended herbicides for groundnut in India are Lasso (alachlor) @ 5 litres/ha or Butachlor @ 0.5 kg ai/ha as pre-emergence sprays within 2 days of sowing or Basalin (Fluchloralin @ 1 kg ai/ha and a post-emergence weedicide) and Fusilade (Fluazifop-p-butyl) @ 0.25 kg ai/ha resulted in yield increase of 71 and 50% respectively over unweeded check.

4.3.5 Pest Management

In India, the major insect pests attacking groundnut are leaf miner (*Aproaerema modicella*), white grub (*Lachnosterna consanguinea*). Red hairy caterpillar (*Amsacta albistriga*), leaf eating noctuid caterpillars like Spodoptera, *Heliothis, Anarsia* and *Plusia*, and sucking pests like aphids, jassids, thrips and white flies. The technology developed to control these pests is given here.

Aphids, jassids, thrips and white flies: Spray Monocrotophos 0.05% or dimethoate 0.05%. Intercropping of mungbean with groundnut should be avoided since thrips, which are the vectors of BND, multiply on this crop. Leafminer: Light traps may be set up to attract and destroy the moths. The economic threshold studies indicated that chemical control should be adopted when the larval population reaches 61-70/100 leaflets. Carbaryl 50 WP 0.2% spray was most economical for controlling this pest.

Red hairy caterpillar: The pupae may be handpicked and destroyed in the field after summer ploughing. Immediately after the receipt of rains, light traps may be set up to attract and destroy the moths. The egg masses and larvae, which can easily be detected, may be collected and destroyed. Before the caterpillars develop hairs on their bodies spraying Monocrotophos 40 EC @ 1,000 ml/ha can effectively control them. A biological control measure through the spray of Nuclear Polyhedrosis Virus (NPV) at the TNAU, Coimbatore was found effective and economical. The virus may be sprayed twice against early instar larvae @ 250 larval equivalents/ha.

Spodoptera, Helicoverpa, Anarsia and Plusia: These pests are nocturnal in habit and hence control measures have to be taken up either during early morning hours or late evening hours, or preferably during night. To control these pests, spray 0.05% Quinalphos @ 400 ml/ha. Other insecticides like Carbaryl 0.2%, Endosulfan 0.04%, and Parathion 0.05% are also effective.

White grub: It is a major problem for groundnut production in the sandy loam soils of north India. Due to white grub menace, the area under groundnut has declined considerably during the last 2 decades in Uttar Pradesh, Rajasthan and Punjab.

The following integrated control was developed to keep the pest under check:

i. The field should be ploughed twice during May-June to expose the beetles present in the soil.

ii. Wherever possible the crop should be sown early (between 10 and 20 June) to reduce the damage.

iii. The adult beetles should be killed by spraying the preferred host trees with 0.2% Carbaryl 50 WP or 50 ml Folithion/Sumithion 50 EC in 100 litres water.

iv. The soil may be drilled with Phorate or Carbofuran or sevidol granules @ 25 kg/ha at least one week before sowing.

v. Wherever the incidence is moderate, a cheap method of seed treatment may be adopted by treating the kernel with chlorpyriphos @ 1.25 litres/100 kg kernels.

According to DARE, Bruchid beetle (*Caryedon serratus*) caused 19-60% losses in groundnut stored for five months. Ten potent bacterial isolates of Bacillus thuringiensis have been identified to tackle this menace. Some potent cultures are BrN1, BINS 5 and SLN 1. The insecticidal crystal protein (ICP), involved in killing the larvae, was isolated from the potent cultures. An insect parasite *Anisopteromalus calandrae* has also been identified. Effective control against leafminer was obtained through spraying Quinalphos 1% at 40 days after sowing. Maximum control of *Spodoptera* could be obtained with Prophenophos at 1% followed by Quinalphos 0.05%. Neem-seed-kernel extract (3 to 4%) and nuclear polyhedrosis virus spray (250 larval equivalents/ ha) efficiently suppressed the sucking and defoliating pests, besides resulted in higher pod yields.

4.3.6 Disease Management

The important diseases occurring on groundnut are *Cercospora* leaf spots (*Cercospora arachidicola* and *Phaeoisariopsis personata*), rust (*Puccinia arachidis*), collar rot, dry-root rot, stem rot (*Sclerotium rolfsii*) and bud necrosis disease (Tomato Spotted Wilt Virus –TSWV). For controlling these diseases the following control measures were developed. *Cercospora* leaf spots and rust: Carbendazim 0.05% + Mancozeb 0.2% should be sprayed 2 or 3 times, at 2-3 weeks interval, starting from 4-5 weeks after planting. In the AICRPO trials this combination has resulted in the best control of both the diseases.

Collar rot and dry root-rot: The seeds should be treated with 5 g Thiram or 3 g of Mancozeb or 2 g Carbendazim/kg kernels. Stem rot: The seeds should be treated as per collar rot. In the soils infected with stem rot causing fungi, frequent disturbances of plant and soil should be avoided to prevent the stems from coming in contact with the organism. In such soils, herbicides may be used to minimize disturbance of the soil.

Bud necrosis disease: Thrips, which transmit the disease, should be controlled by spraying 400 ml Rogor or 360 ml Methyl demeton/ha. Cultural methods like early planting, closer spacing and intercropping with a millet crop are advocated. Growing tolerant varieties like Kadiri 3, ICGS 11 and ICGS 44 will be helpful. However, the maximum reduction of 42.8% of rust was recorded with groundnut + pearl millet intercropping.

4.3.7 Harvesting

The crop must be harvested at the proper time. This will depend on the variety grown and its duration. Normally, the varieties recommended mature in 115-120 days. The symptoms like yellowing and drying of lower leaves, and reddish colour of the kernels after the pods are broken are the visual indicators to decide upon harvesting.

4.3.8 Yield

The crop yield between 2.3 to 2.5t/ ha if proper management practices are followed. Under supplementary irrigation it can yield about 2.8 t/ha.or more.

4.3.9 Drying

Due to high rainfall and humidity in Goa, proper technique of drying and storing groundnuts is to be adopted. Make bundles of 1.5m diameter of harvested plants with pods intact. Keep another heap over it in inverted fashion with pods of both heaps touching together in sun during day time and in the evenings remove the top heap. Repeat this process for three days and two nights. This will help in gradual drying and reduction of moisture level in pods to around 9%. Do not separate the pods from plants immediately after harvest. Thorough drying can be judged by rattling sound when the pods are shaken. When the kernel is pressed between two fingers, it easily spits into two cotyledons. The seed coat comes off easily when rubbed. On the fourth day, separate the pods and sun dry once to store.

4.3.10 Storage

Use HDP bags for storing. Put 250 g of calcium chloride in a muslin cloth placed in a perforated plastic jar, without touching the bottom of the jar vertically in the centre of the storage bag. Store in a moisture proof room. The seed viability would be intact for more than 9-10 months and the seed can be used for sowing in the next season.

Suggestion for the improvement of groundnut production

- Government should provide better methods of drying and storage of seed-pods to the farmers so that they can get better price in future.
- The investment on fertilizers and plant protection has to be increased so that higher productivity is achieved and the cost of production of a unit of pods is reduced.
- Minimum support price should be done optimum to the farmers by Government so that they become interested in cultivating groundnut.
- Marketing facilities should be there near the farmers' field. Market intermediaries should be abolished.

- Farmers should be well acquainted with the new technologies and inventions of agriculture. This can be done through various extension programmes. Result and Method demonstration are very helpful for this purpose.

4.4 Sesame

Sesame (*Sesamum indicum*) also known as 'til' and gingelly, is an important and ancient oil yielding crop. Sesame is regarded as queen of oilseeds by users because of the quality (fatty acid composition) and for its resistance to oxidation and rancidity even when stored at ambient air temperature (Balasubramanian and Palaniappan, 2002). Sesame seed is a rich source of edible oil. Its oil content is generally varies from 46-52%. Its grain may be eaten fried, mixed with sugar or in the form of sweet meats. Sesame oil is used as a cooking-oil in southern India. It is also used for manufacturing perfumed oils and for medicinal purposes. Sesame cake is a rich source of protein, carbohydrates and mineral nutrients such calcium and phosphorus. Sesame cake is mainly used as cattle feed.

4.4.1 Origin and distribution

Archaeological evidences indicate that sesame was cultivated in Palestine and Syria by 3000 BC. In the Indian sub-continent, charred sesame seeds were recorded in Harappa, the oldest archaeological site in the region. In addition largest sesame acreage and diverse forms of sesame are seen in India. On the other side, 25 species of sesame are found in Africa through Africa has limited cultivated varieties. Based on diverse form of cultivated sesame and occurrence of wild sesame (*Sesamum indicum* sp. *malbaricum*), India is now considered as the basic centre of origin. Sesame is basically a crop of the tropics and sub-tropics. Its main distribution is between 25°S and 25°N latitude. The important countries in the world which grow sesame are India, China, Philippines, Burma, Korea, Thailand, Srilanka, Pakistan, Egypt, East Africa, Ethiopia, Mexico, USA and Australia. In India the major sesame growing states are Rajasthan, Uttar Pradesh, Madhya Pradesh, Maharashtra, Andhra Pradesh, Odisha, Karnataka, Gujarat and Tamil Nadu.

4.4.2 Botany

The genus *Sesamum* belongs to the family Pedaliaceae. Sesame is an erect and annual herb. The plant is simple or branched. The leaves are opposite below and alternate above. The flowers zygomorphic, solitary, occasionally two or three together, axillary, short-pedicelled, borne on the upper part of the stem or branches, geneculate, the tube is bent downwards and is dialated above the oblique base. The calyx is small and five sectioned, 0.5-0.6 cm long. The corolla is tubular, campanulate, approximately 3 cm long with applicate margin and slightly bilabiate, the upper lip is entire and the lower one is tripartile. There are four functional stamens and often one is sterile, didynamous. The anthers are sagittate. The ovary is superior, binocular, but may be sub-divided by false internal walls. The fruit is capsule, erect and oblong. The capsule contains numerous small ovate seeds. The testa may be smooth or reticulate and may be white, yellow, reddish brown or black. The seed is dicotyledonous, albuminous and oleaginous. Many classification have been recorded comprises 16 genera and 60 species, of which several can be crossed with *Sesamum indicum* and a few also cultivated for their seeds.

Growth stages

Similar to other annual crop sesame also has different growth phases. In general, only three stages are indicated (i) pre-flowering (establishment to 30 DAS), (ii) Flowering stage (31 to 45 DAS), (iii) Post-flowering stage (46-85 DAS)

In reality, four different stages could be identified for sesame of 85 days duration and they are as follows.

(a) Establishment stage - 0-12 days (12 days)

(b) Vegetative stage - 13-36 days (24 days)

(c) Reproductive stage - 37-60 days (24 days)

(d) Ripening stage - 61-84 days (25 days)

4.4.3 Climate

Sesame is fairly drought-resistant crop. Once established, sesame is capable of withstanding a higher degree of water stress. Sesame is grown in the plains and at elevations up to 1200m. It cannot withstand frost, continued heavy rain or prolonged drought. Sesame produces an excellent crop with a rainfall of 500-650 mm. Heavy rain at flowering will drastically reduce yield and if cloudy weather persists for longer period at this time, yield can be exiguous. Similarly, rainfall during harvest also reduces the yield by increasing susceptibility to diseases. Sesame requires fairly hot condition during growth to produce higher yields.

4.4.4 Land requirements

Select lands on which sesame was not grown in the previous season unless it was of the same variety and grown from the certified seeds. In addition, the soil of selected fields should be rich in organic matter, free of weeds, well aerated and well drained. Soils should be free from excessive salts with near neutral in reaction.

4.4.5 Isolation requirements

Sesame is mainly a self-pollinated crop, but some cross-pollination by insects occurs. Cross pollination upto 5% has been recorded. For pure seed production, the seed fields must be isolated from other sesame fields by hundred meters for foundation seed class and fifty meters for certified seed class (Agrawal, 1980).

4.4.6 Variety

The varieties differ not only in respect of the season of growth and the time of maturity but also in the number of capsules formed in the leaf axils, the habit of growth (Single stemmed or branched) and the size, colour and oil content of the seeds.

In India, each major sesame growing state releases its own sesame varieties suitable to their agro-climatic conditions. Important varieties suitable for different states are given below:

State	Variety
Andhra Pradesh	Gowri (*Kharif* and *Rabi*), T85 (*Kharif*)
Gujarat	Mrugi (*Kharif*), Purva (Semi *Rabi*), Patan 64 (*Kharif*) and Patan 65 (*Kharif*)
Madhya Pradesh	G5 (*Kharif*), T4 (*Kharif*), G35 (*Kharif*), No.32 (*Kharif*), No. 128 (*Kharif*)
Maharashtra	No.128 (*Kharif*), Chanda-8 (Semi *Rabi*), N 58-2 (*Kharif*)
Odisha	Vinayak (*Kharif/Rabi*)
Rajasthan	Pratap (*kharif*), RT 125 (*kharif*), Type 13 (*kharif*)
Tamil Nadu	RT 125 (*kharif*), TMV3 (*kharif*), TMV4 (Summer), TMV 6 (Summer), TMV (*Rabi*), VRI1 (Summer)
West Bengal	B-67

4.4.7 Agronomic management

4.4.7.1 Land preparation

For sesame cultivation, medium fertile soil, with high drainage and a mixture of sand and clay texture are highly suitable. Generally, sandy clay or sandy loam soils are preferred for sesame. The soil should have a pH range between 5.3 to 8.0. Land preparation is very important for sesame because sesame seeds are very small and sensitive to tilth of the soil. Sesame seed germination is positively correlated with soil tilth. The germination of seed and establishment is better in fine tilth soil. The field should be free from stubbles and weeds. The field should be well leveled.

One deep ploughing and two to three harrowing followed by leveling are sufficient to prepare field to the desired tilth for seed production of sesame.

4.4.7.2 Sowing parameters

Cultivation is done with nucleus /breeder / foundation seeds from source approved by seed certification agency.

In India, sesame yield is higher in summer irrigated crop than the monsoon crop. Reduction in monsoon yield is mainly due to heavy rain received during the crop growth period. In most of the states in India, sesame is being grown during *kharif* season as monsoon crop. In Madhya Pradesh, Maharashtra, Andhra Pradesh, Tamil Nadu, Odisha and Karnataka it is grown both in *kharif* and *rabi* season. It is also grown in summer in parts of Odisha and Tamil Nadu under irrigation.

Seed crop must be sown in lines with the help of seed-drill. The spacing for row to row is 30 to 45 cm and plant to plant is 15-22 cm. The depth of seeding should be 2.5 to 3.5 cm. The seed rate varies from 3-5 kg/ha. The seed should be treated with mancozeb @ 3 g/kg of seed to control seed-borne disease. For bio-seed treatment, 1.0 kg of seeds and 4.0g of *Trichoderma viridae* powder, which contains active fungi is to be mixed and kept under shade for an hour before sowing.

4.4.7.3 Nutrient management

In reality, sesame crop is raised by small and marginal farmers on relatively poor soils without any proper nutrient management. However, for satisfactory yield, adequate nutrients are to be provided. The crop responds well to organic manures. On low fertility lands, apply 34 kg nitrogen, 17 kg phosphorus and 34 kg potassium per hectare. On fertile land the crop can be grown even without fertilizer, or apply 30 kg nitrogen and 20 kg phosphorus per hectare. It is desirable to meet this fertilizer needs through application of FYM or compost. The results from West Bengal indicated the superiority of combined application of 4 tonnes of FYM/ha over 45 kg N + 22.5 kg P_2O_5/ha for sesame crop. Mondal *et al.* (2008) reported that maximum seed yield of sesame (670.4 and 744 kg/ha) was obtained when rice and sesame in rice-lathyrus-sesame sequence received 50% recommended dose of N, P & K in addition 10 tonnes FYM/ha and green manuring to rice.

Recommended dose of N, P and K for sesame in different states under irrigated condition

State	N, P & K (kg/ha)
Andhra Pradesh	40-20-20
Assam	30-20-20
West Bengal	40-20-20
Bihar	30-25-0
Gujarat	50-25-25
Himachal Pradesh	60-40-20
Odisha	30-20-20
Karnataka	40-20-20
Maharashtra	50-0-20
Rajasthan	50-25-25
Tamil Nadu	40-25-25
Uttar Pradesh	30-20-25
Madhya Pradesh	30-30-15
Jammu and Kashmir	20-10-0

Source: Pasricha *et al.* (1991) and P.S. Rathore (2002)

4.4.7.4 Irrigation management

Irrigation requirement of sesame varies with type of varieties, soil type and season of cultivation. The water requirement of sesame is varied from 250 mm to 350 mm. The *kharif* and semi-*rabi* crops are entirely rainfed, whereas summer crop is grown under irrigation. The yield of *kharif* crop is poor, whereas those of the semi-rabi and summer crops are high as they are grown in rich soils and under better management. Irrigation few days after sowing (if soil moisture is less), when plants are 15-20 cm high (pre-flowering), and during flowering and maturation of seed are important for obtaining increased seed yields. Irrigation may also be scheduled at 0.6 IW/CPE ratio in vegetative stage followed by 0.4 IW/CPE ratio at reproductive stage.

4.4.7.5 Weed management

Weed infestation is more in *kharif* season sesame as compared to *rabi* and summer seasons. Due to slow growth of sesame in the early stage, weeds can be controlled properly, otherwise the growth and seed yield will be hampered. Two to three hoeing and weeding are necessary to control weeds of sesame. The first hoeing should be done prior to irrigation, when the plants are 15-20 cm high. Subsequent hoeing may be done as per requirements. Integrated weed management practices recommended for sesame for better weed control and yield, i.e. pre-emergence application of fluchloralin @ 1 kg/ha + one hoeing and hand weeding at 30 DAS.

4.4.7.6 Plant protection measures

Insect Pest

To control caterpillers, gall fly and leaf roller, spray monocrotophos @ 625 ml/ ha or soil application of Malathion dust (5%) @ 25 kg/ha.

Diseases

Powdery mildew and black stem wilt are generally infested the sesame crop. General recommendation to control sesame diseases are given below.

(i) Dry seed treatment with thiram / carbendazim / mancozeb @ 2.5 g/kg seed or bio-treatment with *Trichoderma viridae* @ 4 g/kg of seed.

(ii) Spraying 500 g carbendazim or 1000 g mancozeb/ha at 35 DAS along with insecticide.

(iii) Spraying of wettable sulphur at 2.5 kg/ha along with insecticide at 50 days after sowing of sesame.

4.4.7.7 Roguing

Roguing should be done before flowering, after flowering and before harvesting. All the off-type plants and diseased plants affected by rosette and leaf spot blight should be removed from the field from time to time as required to ensure pure seed production.

4.4.7.8 Harvesting and threshing

The sesame crop is harvested when the leaves, stems and capsule begin to turn yellow and the lower leaves start shedding. To prevent the shedding of seeds, the crop should not be allowed to become dead ripe in the field. The ripe plants are cut, carried to the threshing floor, stacked for a week in the sun with the cut ends downwards and then well shaken or beaten to take out the grains from the dry capsules. Then winnowing and cleaning has been done. Dry the seeds in the sun for 3 days to 5% seed moisture.

4.4.7.9 Yield

The average seed yield of *kharif* sesame is 3 to 5 q/ha and the average yield of semi-*rabi* and the irrigated summer crop are 5 to 6 q/ha.

Important hints for obtaining high yields -

(i) A suitable variety must be chosen for the tract and the season.

(ii) The crop must be sown on time in lines.

(iii) The application of fertilizers, inter-cultivation and plant protection measures must be taken up at the appropriate time.

(iv) Harvesting should be done after the crop matures, but must not be delayed up to the stage when the capsules dehisce and shed seeds.

References

Agrawal, R.L. (1980). Seed Technology. Oxford and IBH Publishing Co., New Delhi.

Balasubramanian, T.N. and Palaniappan, S.P. (2002). Sesame. In: Techniques and Management of Field Crop Production. Published by Agrobios (India), Jodhpur : 178-196.

Mondal, S.S., Jana, P. and Chandra, P. (2008). Integrated nutrient management for sustaining productivity and soil fertility build-up in rice (*Oryza sativa*) – Lathyrus (*Lathyrus sativus*) – Sesame (*Sesamum indicum*) cropping system. Indian J. Agri. Sci. 78(6): 534-536.

Pasricha, N.S., Bahl, G.S., Aulakh, M.S. and Dhillon, K.S. (1991). In: Fertilizer-use Research in oilseed and Pulse crops in India. ICAR.

Rathore, R. (2002). Techniques and management of field crop production. Published by Agrobios (India): 308-316.

4.5 Linseed

Linseed (*Linum usitatissimum*) is a very important among oilseeds owing to its various uses and special qualities. It is grown for both oil and fibre (flax). In India, it is grown mainly for seed used for extracting oil. The oil content of the seed varies from 35-47%. Linseed oil is an excellent drying oil used in manufacturing paints and varnishing, oil cloth, water-proof fabrics and linoleum and as edible oil in some areas. Linseed cake is a very good organic manure and animal feed. The fiber of linseed straw is known as lignin. Linseed is used in making paper and plastics.

4.5.1 Origin and distribution

Linseed has been grown from ancient times for fiber (flax) and for its seed which is rich in oil. The oldest region of cultivated flax is reported to be in Asia and on the Mediterranean coast. Vavilov (1926) reported South-West Asia namely India and Afganistan as centre of origin of linseed as oilseed purpose. Linseed is extensively grown in the countries of the temperate zone as well as in those of the tropical zone. The major linseed growing countries are Argentina, India, USA, Canada, Pakistan, Australia, Germany, Russia and China. In India, it is mainly grown in Madhya Pradesh, Uttar Pradesh, Bihar and Maharashtra.

4.5.2 Botany

The cultivated linseed (*Linum usitatissimum,* n=15) belongs to family Linaceae. Linseed is an annual herb with a thin, erect and wiry stem, about 60-100 cm high. The fibre-type cultivars are generally slender, tall, non-tillering and slightly branched. The cultivars for seed purposes are usually smaller, much branched and profusely tillering. The small flowers with bright blue colour but sometimes pink or white petals grow on terminal panicles. The flowers are self-pollinated, although, cross fertilization is possible. The fruit is a round capsule divided into five chambers, each containing two yellow or brown seeds.

Classification

Linseed belong to the family Linaceae and the genus *Linum* which has 100 species. Riccharia (1962) classified the 26 varieties of Indian linseed on the basis of seed colour and size.

1. Based on seed colour

(i) Yellow coloured

(ii) Brown coloured

(iii) Fawn coloured

2. Based on seed size

(i) Small seeded

(ii) Medium seeded

(iii) Bold seeded

4.5.3 Climate

Linseed is grown in all the climatic conditions like tropical, sub-tropical and temperate from 11-30°N Latitude. Its cultivation is confined to low elevation, but it can be successfully grown upto 770 meters. Area with the annual rainfall ranging from 45-75 cm is best suited for its cultivation. The linseed as seed crop does well under moderate cold, but the fibre crop grows best in cool moist climates.

4.5.4 Variety

In India, the varieties are of two types as given below :

(i) Peninsular types which have a deep root-system.

(ii) Alluvial types which are shallow-rooted and are characterized by profuse tillering, varieties differ in maturity, growth habit and seed size and colour.

Selection of linseed variety is an important factor to harvest high yield. The improved varieties of linseed with their characteristics and area of adaptability are given below in Table 4.1.

Table 4.1: Improved varieties of linseed cultivated in different states

Variety	Duration (Days)	Average yield (q/ha)	Oil content (%)	States
K-2	175	11.0	46	Haryana, Punjab, Himachal Pradesh
Hira	135	12.0	43	Uttar Pradesh
Mukta	130	12.0	45	Uttar Pradesh
Jawahar 17	120	8.0	43	Madhya Pradesh
Neelam	145	13.0	43	Uttar Pradesh, West Bengal
Himalini	165	13.0	42	Himachal Pradesh, Punjab, Rajasthan and Haryana
B-67	125	13.0	42	West Bengal
Garima	125	15.0	42	Uttar Pradesh
Sweta	130	12.0	44	Uttar Pradesh
Pusa 2	150	13.0	43	Major Linseed growing areas
Gaurav	137	10.0	43	Good for fiber
Shubhra	130	12.0	45	Uttar Pradesh

4.5.5 Isolation requirements

Linseed is normally a self-pollinated crop. Natural cross-pollination upto 3% mainly due to insects has been recorded. For pure seed production, it is desirable to isolate the varieties at least 10 to 300 meters apart. Distance can be reduced if another tall crop of linseed is planted between adjacent varieties (Agrawal, 1980).

4.5.6 Agronomic management

4.5.6.1 Selection of field

For seed production of linseed, select fields in which linseed crop was not cultivated in previous season, unless the variety grown was the same and the field approved by the certification agency for varietal purity.

4.5.6.2 Soil and land preparation

Linseed can be grown on different kind of soils, except sandy and badly drained heavy clay soils, but does best on clay loam soil. It can grow well on the deep clayey black soils of central and peninsular India and on the alluvial loams of the Indo-Gangetic plains. The field should be well drained, free of weeds and stubbles. The soil should have a good water-holding capacity. In general, one ploughing, two to three harrowing followed by leveling, is sufficient to desired tilth. Frequent ploughing are given to retain moisture in the soil and to check weeds in the field during the rainy season. A fine weed-free seed-bed is prepared for sowing of linseed for seed production. Chlorpyriphos 5% dust @ 25 kg/ha should be mixed within soil during last ploughing before sowing the seed to protect the seedlings against damage by which ants and cutworms.

4.5.6.3 Sowing parameters

The cultivation is done with nucleus / breeder / foundation seeds from source approved by seed certification agency.

The main season for sowing of linseed is winter. Sowing time varies from the first fortnight of October to Mid-November in various states, depending upon the cessation of the rains. Mainly time of sowing varies on climatic conditions and availability of moisture under rainfed situation. In Jammu & Kashmir crop is sown in February-March. The linseed crop may be sown by broadcast or in lines 20 to 30 cm apart and plant to plant is 5 to 6 cm. The depth of sowing should be 5 cm. In case of broadcast sowing which includes *utera* sowing also, the seed rate is 40 kg/ha. For line sowing, the rate varies from 20-30 kg/ha depending upon seed size. Thin sowing of crop gives a better number of tillers, but for fiber and seed production, thick sowing is desirable.

4.5.6.4 Nutrient management

The application of 5-7 t FYM per hectare not only improves nutrient availability, but also soil physical condition, particularly water retaining capacity. Crop is grown both under rainfed and irrigated conditions. The application of fertilizers and irrigation increases the yield of the crop. Under rainfed conditions, application of 30 kg N and 15 kg P_2O_5/ha is beneficial. Under irrigated conditions, nitrogen is applied in two splits-half the dose as basal and the other half at the first irrigation (40 to 50 days after sowing). Singh and Verma (1997) recommended 90 kg N, 45 kg P_2O_5 and 30 kg K_2O/ha for light textured soils of UP.

4.5.6.5 Irrigation management

Under rainfed situation the seed yield of linseed (as *utera* crop) was increased significantly with application of 10:20:10 kg N, P_2O_5 and K_2O/ha to linseed after preceding rice crop (Mondal *et al.*, 2004). A similar finding was also reported by Kurchania *et al.* (2002). Crop is mainly grown under rainfed condition, however, it responds well to irrigation. Irrigation during early growth period and flowering time greatly enhances seed yield. Singh and Verma (1997) reported seed yield with no irrigation, one irrigation at branching and one irrigation at capsule initiation as 6.79, 8.10 and 7.79 q/ha respectively. Dutta *et al.* (1995) obtained 11.68 q/ha at 0.8 IW/CPE as compared to 10.75 q/ha at 0.4 IW/CPE ratio, while under no irrigation yield was 9.88 q/ha. Singh and Verma (1997) reported seed yield with no irrigation, one irrigation at branching and one irrigation at capsule initiation as 6.79, 8.10 and 7.7 q/ha respectively. Thus two irrigations first at 35 days after sowing (seedling development) and second at 65 days after sowing (flower initiation) augmented the yield to great extent.

Froment *et al.* (2000) reported that at 50% capsule formation of linseed, crops receiving extra N had darker leaves. Capsule number, seed number and seed weight per pod significantly increased compared to the control when additional N was applied.

Mondal *et al.* (2003) reported higher yield of *utera* crop when P dose of rice was omitted in the rice season and was given instead to the *utera* crop in addition to its recommended dose of fertilizers.

4.5.6.6 Weed management

The rainfed crop generally requires no intercultural operations. But when the crop is irrigated, one weeding becomes essential. Linseed plant is short-stature, and grows slowly in the beginning and has less number and area of leaves. Therefore, it cannot compete with weeds. Generally, 30-40% reduction in grain yield due to weed infestation is also observed in linseed crop. One weeding during early stage of growth and one after first irrigation is necessary. The common broad leaf weeds of linseed are *Chenopodium album, Carthamus oxycantha*, *Melitotus alba* and *Anagallis arvensis* and grassy weeds are *Avena ludoviciana, Phalaris minor, Cyperus rotundus,* and *Cynodon dactylon.* Weed can be controlled with the application of Fluchloralin @ 1 kg/ha as pre-plant incorporation or Isoproturon @ 0.75-1.0 kg/ha as post-emergence, one month after sowing. Mondal *et al.* (2005) reported maximum yield of rice as well as linseed as *utera* crop (4673 kg/ha and 664 kg/ha respectively) was obtained under weed-free check where hand weeding was done twice at 20 and 40 days after sowing, followed by butachlor at 1.5 kg a.i. ha + one hand weeding at 40 DAT treatment (4605 kg/ha and 619 kg/ha of rice and linseed respectively).

4.5.7 Roguing

All the off-type plants distinguishably on the basis of plant character must be removed before to ensure pure seed production of linseed. The diseased plants affected by wilt and rots should be removed from time to time in the field as required.

4.5.8 Plant protection measures

Insect-pest

The important insect-pest and diseases infesting linseed and their control measures are given below. To control linseed bud fly spray phosphamidon 0.03% at 15 days interval. Leaf minor can be controlled by dusting with 10% methyl parathion @ 25 kg/ha or spray 0.04% monocrotophos 36 EC. To control caterpillar, spray 1.25 lit Thiodan 35 EC in 500-700 litres of water.

Disease

Blight (*Alternaria* sp.) can be controlled by early sowing, crop rotation, seed treatment with bavistin @ 2 g/kg of seed and spray 0.25% mancozeb. Rust (*Melanospora lini*) can be controlled by spraying Dithane Z-78 @ 2 kg/ha in 500-750 lit of water or dusting 20 kg/ha sulphur dust. Powdery mildew (*Oidium lini*) can be controlled by spraying wettable sulphur compounds as Sulfex @ 2-3 kg/ha in 500-750 lit of water. Repeat spray after 15 days.

4.5.9 Harvesting and threshing

The linseed crop starts maturing by the middle of February, depending on winter spread and sowing time. The maturity of linseed crop can be judged by colour of leaves, stem and grain. Plants turn golden yellow when the crop is mature and ready for harvesting. Harvesting is done when the crop is dead ripe with a sickle or by uprooting the plants. When the fibre of linseed is also desired along with seed, the harvesting of the crop is done at the stage of capsule maturity even when the crop is slight green. For oil crops, cutting is preferred and for fibre, uprooting is better because of more fibre length. After harvesting, produce is kept for sun-drying. Grains are separated after 2-3 days by beating plants with stick or by trampling it under the feet of bullocks. Seed is separated from chaff by winnowing.

4.5.10 Yield

The average yield of a pure linseed crop varies from 12-15 q/ha under irrigated condition and 210 to 450 kg per hectare under rainfed condition. The crop in northern India generally given higher seed yield than in central and peninsular India.

4.5.11 Storage

Seeds are dried in sun till moisture is 8-10%. After drying, seed may be stored in a dry place.

Important hints for getting maximum yield -

(i) Select the recommended varieties.

(ii) Clean, fine, weed free seed-bed should be prepared by ploughing.

(iii) Carbofuran granules @ 25-30 kg/ha be mixed with the soil at the last ploughing before sowing.

(iv) Early sowing escapes the attack of many diseases and pests under *utera*, sowing be done at the dough stages of rice.

(v) The recommended doses of N, P and K should be applied to the sole and *utera* crop for seed production.

(vi) Use optimum seed rate of 30-40 kg/ha. A closer row spacing of 20-30 cm and 40 kg seed per hectare is followed for *utera* sowing.

(vii) Timely plant protection measures should be taken.

References

Agrawal, R.L. (1980) Technology. Oxford and IBH Publishing co., New Delhi.

Dutta, H.K., Ram, D.S., Rao, M. and Singh, H. (1995). Response of linseed (*Linum usitatissimum*) to irrigation and nitrogen. Indian Journal of Agronomy.

Froment, M.A., Turley, D. and Collings, L.V. (2000). Effect of nutrition on growth and oil quality in linseed. Tests of Agrochemical Cultivars 21: 29-30.

Kurchania, S.P., Sharma, R.S., Mathew, R. and Dwivedi, R.K. (2002). Performance of utera crops in relation to growth duration of rice (Oryza sativa) varieties under rainfed condition. In: Extended Summaries of Second International Agronomy Congress on Balancing Food and Environmental Security - A Continuing Challenge. 26-30 November, 2002 at New Delhi, pp. 1079-1080.

Mondal, S.S., Acharya, D., Saha, M, and Sounda, G. (2005). Evaluation of efficiency of weed management in rice-based paira cropping system. In: Proceedings of The 20th Asian Pacific Weed Science Society Conference. 7-11 November 2005 at Ho Chi Min City Vietnam, pp. 564-567.

Mondal, S. S., Ghosh, A., Acharya, A. and Sarkar, A. (2004). Performance of linseed as utera; crop as influenced by varietal duration of rice under rainfed low land Ecosystem. Journal of Interacademicia 8(3): 364-370.

Mondal, S.S., Ghosh, Arup and Acharya, D. (2003). Developments of Agro techniques for sustainable productivity of rice-based utera cropping system. NATP Technology Bulletin No. 1, September 2003, BCKV, Mohanpur, W.B.

Riccharia, R.H. (1962). Linseed: Indian Central Oilseed Committee, Hyderabad. pp. 155.
Singh, N.B. and Verma, K.K. (1997). Response of linseed (Linum usitatistimum) to varying irrigation and fertility levels. Indian Journal of Agronomy 42(4): 696-698.
Vavilov, N.I. (1926). Studies on the origin of cultivated plants. Bulletin of applied Botany and Plant Breeding. 16(2).

4.6 Soybean

Soybean (*Glycine max.* L. merril) with its 40% high biological value protein and 20% oil, is considered highly nutritive crop. It is the rich source of calcium and vitamin A. Soybean protein has 5% lycine, which lacks in most of the cereals. It contains highest quantity of thiamine and riboflavin among different legumes. Soybean contains less starch, thus it is better for diabetic patient. The soybean oil is used as cooking medium and also for making Vanaspati ghee. It has great importance for industrial uses like soya flour, soya cake, biscuits, soya milk, varnish and paints. Soybean also enriches soil by fixing atmospheric nitrogen. Soybean plant is used as fodder and cake as an excellent concentrate for livestock and it also enriches soil by fixing atmospheric nitrogen. Approximately 85% of soybean produced is utilized for oil extraction, 10% for seed and 5% for food. To expedite the higher production potential of the new strains, the programme of seed production requires strengthening and its subsequent development to a stage where participation of private enterprises in seed business becomes a practical and profitable position. The Government can also support the private and voluntary agencies in the production and supply of improved quality of seeds for higher total production of soybean.

4.6.1 Origin and distribution

The genus *Glycine* is divided into three sub-genera with *Glycine max*, the soybean, *Glycine ussiriensis* and *Glycine gracilis* in the sub-genus *Soja*, which is its centre of diversity in east China, Korea, Taiwan and Japan. Of the other two sub-genera one is endemic in Australia and the other in Africa and India. The three species in sub-genus *Soja* are interfertile and some authorities suggest that they constitute a single species. *Glycine ussiriensis* is a wild species and is the putative ancestor of soybeans.

Based on historical and geographical evidences, North-eastern China is generally regarded as the origin place of soybean. Soybean has been cultivated from pre-historic time in China. It was introduced in USA during 1804 and in India during 1880 (Rathore, 2002). The other important countries producing soybean are Brazil and Argentina. In India, it has been cultivated in Madhya Pradesh, Uttar Pradesh, Rajasthan and Maharastra.

Classification

Soybean has been classified in three different groups on the basis of colour, shape and size of seed, shape of pods and maturity period.

A. Manchurian Classification – It is based on the seed colour.

(i) Yellow group – Seeds of this group are yellow with (a) light hilum, (b) Golden hilum (c) Dark brown hilum

(ii) Black group – Seeds are of black colour with (a) small seed (b) large seed (c) Flat seed

(iii) Green group – Epidermis of seed is green with (i) yellow embryo (ii) Green embryo

B. Martin Classification – It is based on the shape and size of seed.

(a) *Soja elliptica* with elliptical seed

(b) *Soja compressa* with compressed seed

(c) *Soja spherical* with rounded seed

C. Hertz classified the soybean on the basis of pods – It is based on the shape and size of pod.

(d) Soja tumida with swollen pods (e) *Soja platycarpa* with flattened pods.

4.6.2 Botany

The cultivated soybean under the Family: Leguminosae; sub-family Papilionaceae, belongs to genus *Glycine*. (Chromosome No. 2n = 40). *Glycine max* (L.) Merrill is a hairy annual, erect, bushy with a height of 0.5 to 2 metre. It has a tap root system with large number of secondary and adventitious roots. Roots may go as deep as 150 cm but most of the active root system is restricted to only 15 to 20 cm from soil surface. Root system is extensively nodulated. The trifoliate leaves develop on main stem and branches. These are alternatively arranged and prophylls present on the base of each lateral branches. The flowers are in short axillary or sometimes terminal, racemes on the main stem and branches, each raceme with 2-35 small, white or pale purple flowers. Flowers are normally self-pollinated. Pods are small, straight flattened or cylindrical, occurs in groups of 3 to 15. Their colour and size varies, but commonly they contain two to four seeds and are 3-7 cm long, hairy, light brown when mature, and slightly constricted between the seeds. The seeds vary greatly in size, shape and colour, but are most often round and yellow or yellowish green, brown or black.

4.6.3 Climate

Soybean can be grown in warm and moist climate. A temperature of 25-30°C is optimum for the growth and development of soybean crop. Lower temperature tends to delay the flowering. If temperature is below 10°C or above 38°C the crop growth is retarded. Crop can be grown in areas receiving 600-650 mm rainfall. Rains just after sowing is harmful for its germination as well as rainfall during maturity deteriorate the grain quality to a great extent. Cloudy weather prolongs vegetative phase.

4.6.4 Agronomic management

4.6.4.1 Selection of land

Select a field in which soybean was not grown in the previous season, unless the variety grown was the same and approved by the certified agency for varietal purity. The field should be well drained and the soil neither too acidic nor too alkaline.

4.6.4.2 Isolation requirements

Soybean is normally a self-pollinated crop. The dehiscence of anthers takes place in the bud itself before the opening of the flower and hence, normally self pollination takes place. Cross-pollination by insects is usually less than 1%. An isolation of 3 m from other fields of soybean is sufficient to maintain genetic purity of seed.

4.6.4.3 Soil and land preparation

Soybean can be grown on variety of soils ranging from light to black cotton soils. Soybean requires a well drained, neutral, fertile loams soil for its optimum growth and yield. Soil should be loose and well aerated. Highly compacted soils are harmful for root nodule development. Water-logging is also harmful for soybean crop. Soybean requires a weed free, well leveled and pulverized seedbed like other legumes. One deep ploughing followed by two harrowing and planking is sufficient to prepare a good field for sowing of soybean. To protect the crop from soil-borne insects specially termites 25 kg/ha chlorpyriphos dust 4% should be mixed into the soil just before last harrowing. Apply chlorpyriphos @ 3 ml/lit of water for controlling white ant.

4.6.5 Sowing parameters

Cultivation is done with neucleus / breeder / foundation seeds from source approved by seed certifying agency.

Seed treatment and *Rhizobium* inoculation

Various fungi may attack, directly on seed and seedling also. The important disease causing damage in soybean are *Rhizoctonia, Fusarium* and *Phythium* sp. To reduce the damage of these fungi seed should be treated with fungicide like Thiram/Bavistin/Mancozeb @ 3 g/kg of seed. After treating the seed with those fungicides, appropriate *Rhizobium* culture should be used to improved soil fertility.

Seed inoculation with *Rhizobium* culture for soybean which is grown first time in the field is positively necessary. *Rhizobium* inoculation of seed especially in new soils may increase seed yield and protein content of seed. Two types of soybean culture are available, namely, peat culture and soil culture. The inoculation of seeds should be done in the shade. Increase the quantity of culture in double if the crop is raised on soils new to the soybean crop. Sowing of soybean seed soon after onset of rains is best as compared its earlier or later sowing. First fortnight of July is the most appropriate time for sowing soybean seeds. The optimum sowing time of soybean in different zones are given below (Rathore, 2002).

Table : Optimum sowing time of soybean in different zones

Zone	Sowing time
Northern Hill Zone	Early June to mid July
Northern Plain zone	Last week of June to Mid July
Central zone	Mid June to Mid July
Southern zone	Mid June to mid July
Eastern zone	Late June to Early July

The soybean as seed crop must be sown in rows with seed drill or hand tyne or by plough in furrows. The depth of sowing should be 2-3 cm in fields where soil moisture conditions are optimum for germination of seed and growth of the crop. Sowing upto 4-5 cm depth can be done on light soils.

Sowing may be done in rows at a distance of 45 to 60 cm between rows and plant to plant spacing 4 to 5 cm. Seed rate of about 65-70 kg per hectare is sufficient for seed production of soybean.

4.6.5.1 Nutrient management

Soybean is a leguminous crop, and its requirement of nitrogen is substantially fulfilled by atmospheric nitrogen fixing through *Rhizobium.* Seed treatment with appropriate *Rhizobium* culture not only enhances yield but enriches the soil also. Integrated nutrient management with FYM @ 10-15 t/ha along with 20 kg N/ha as starter dose enhances yield of soybean. Soybean needs large quantity of phosphorus, particularly at the beginning of pod formation till 10 days prior to seed development. So for seed production of soybean, apply 20 to 25 kg N/ha, 80-100 kg P_2O_5/ha and 30 to 40 kg K_2O/ha at the time of sowing.

Phosphorus and potassium requirement of crop should be based on soil test value. In absence a soil test, a dose of 50-60 kg P_2O_5/ha is generally recommended for the crop. This not only increases test weight but also yield (Rajput *et al.,* 1991). Potassium requirement of soybean is also high. The maximum K is required for vegetative development of plant. In absence of soil test, a general dose 20 kg K_2O/ha is recommended for the crop. In zinc deficient soils, one foliar spray of zinc sulphate along with 2.5 kg lime in 1000 lit of water on soybean crop is beneficial. The soybean-sesame intercropping revealed that 80 kg K_2O/ha plus 20 or 40 kg N/ha significantly increased the yield parameters and seed yield of both soybean and sesame under sole and intercropping system (Mondal and Pramanik, 1996). Mondal *et al.* (2004) revealed that the interaction effect of the seed yield of soybean and sesame as sole and intercrops was influenced significantly in the summer and rainy seasons with application of K. Both in summer and rainy seasons, the maximum seed yield of soybean and sesame as intercrops were obtained when crops were fertilized with 80 kg K_2O and 40 kg N/ha.

4.6.5.2 Water management

Irrigation should be given to the soybean crop as and when required. Flowering and pod filling stage are the most critical stages of which moisture stress severely damages seed yield. Moisture stress at flowering and pod formation result in flower and pod dropping. There should be adequate moisture in soil, particularly during flowering, seed development and the maturation stage to obtain high seed yields of soybean. Thus in case of moisture stress, one to two irrigations at flowering and pod filling stage is beneficial in realizing good yield. The last irrigation should be given at least 30 days before harvest.

4.6.5.3 Weed management

Soybean is generally infested with many grassy as well as broad leaf weeds as a rainy season crop. The major grassy weeds of soybean are *Echinochloa colona, Cyperus rotundus, Cynodon dactylon* and broad leaf weeds are *Celosia argentia, Digitaria* sp., *Phyllanthus niruri, Trianthema portulacastrum, Vernonia* sp. and *Tribulus terrestris*. Weeds in soybean field can be controlled by using khurpi or hand-hoeing. Generally, two weedings, one at 20-25 days after sowing and second at 35-40 days after sowing keep most of the weeds under control. So, weed control by hand weeding or with herbicides such as fluchloralin 1 kg/ha or clomozen @ 1.5 kg/ha or both, should be adapted, if required (Joshi and Billore, 1998).

4.6.6 Roguing

Rogue the off-type plants and diseased plants affected by yellow mosaic virus and soybean mosaic virus as soon as they appear, so as to check further spread upto first 2-3 weeks. Continue removal of plants affected by soybean mosaic from the seed field upto last stage. It enhances the seed purity. So, for seed crop rouging is done in the following stages :

(i) At flowering stage, remove off-type plants on the basis of plant to plant characteristics and flower colour.

(ii) At maturity stage, rouging should be done on the basis of pod characters of soybean.

4.6.7 Variety

A number of promising varieties have been evolved in various soybean growing states for commercial cultivation. Characteristics of some of the important varieties of soybean is given below.

Variety	Duration (days)	Plant and seed charaeteristics	Average yield (q/ha)	Oil content (%)	Place of adaptation
Lee	110-115	Plants are shorter erect and branched; pods are in cluster with 3-4 seeds. Flower colour is purple. Seeds are light yellow in colour with black hilum.	20-25	23-25	Himachal Pradesh Drinaga Valley Kumaon Hills of UP
Bragg	120-125	The plants have determinate habit of growth with green stem. Flowers are white and the seeds are bold, yellow with black hilum.	25-30	20	Central zone Northern Hill and Plain zone
Clark	95-100	Plants have indeterminate habit of growth with purple stem. Flower colour is purple seeds are yellow with black hilum.	25-30	18-20	For spring/summer crop to be sown in February only
Improved Palican	110-120	The flowers are purple and seeds are light yellow. It is moderately resistant to yellow mosaic.	20-22	20-22	Southern zone
Punjab 1	120-130	The leaves and pods are covered with brownish hairs. Flowers are purple and seeds are yellow. The pods contain 2-3 seeds.	20-25	20-22	Punjab, Delhi, Haryana, UP, MP, parts of Rajasthan
Shilajeet (UPSM-19)	105	It is an early variety. The flowers colour is purple. Seed is bold.	20-25	20-22	Gujarat, Rajasthan, UP and MP
Ankur (UPSM 38)	125	The seeds are small and yellow in colour with light brown hilum.	25-30	-	Northern Plain zone
Alankar	120	The plants are medium in height with white flower, the seeds are round, yellow with light black hilum.	-	-	-
PK-262	130	The plant is resistant to lodging, It has bold yellow seeds.	30-35	22	UP, Terai region, MP
PK 427	125	It has bold seed. It contains 40.84% protein. It was developed at Pantnagar.	26-30	19.9	UP, Terai region, MP
JS 2	105-110	Bold light yellow seeds.	20-25	-	Central Zone
K30	125-130	Seed is black.	15-18	20	-
Durga	105-110	The variety was developed at Jabalpur.	25-30	-	MP

Yield potential of varieties differ from one agro-climatic region to another. The varieties found suitable for particular region are given below.

Varieties of Soybean suitable for different regions

Regions / Zones	Variety recommended
North Hill Zones	Bragg, Shilajeet
North Plain Zones	Bragg, Alankar, PK 327, PK 262, PK 308, PK 416
Central Zones	Jawahar, Gaurav (JS-72-44), Durga (JS-72-280), Bragg, Ankur, MACS-13
Southern Zone	KHSB-2, Hardee, Improved Pelicon

Plant protection measures

To prevent the spread of yellow mosaic disease in soybean field, spray a mixture of malathion (0.1%) and metasystox (0.1%) at 10-15 days interval. Growing of resistant varieties 'Shilajeet' reduces the damage. Inspect the crop regularly, remove the infected plants and burn them. Spraying of metasystox 1 L/ha in 500-600 water of water is done at 10-15 days interval to control the disease. To control the spread of rust, spray 2.5 kg Dithane Z-78 per hectare once on the 60th day and repeat the spray once more at 10 days later. Hairy caterpillar causes heavy damage in soybean. This insect can be controlled by spraying monocrotophos 0.4%.

4.6.8 Harvesting and threshing

For seed production, the time of harvest and method of threshing is most important for maintaining seed quality. To retain higher seed germination, seed crop should be harvested as soon as it is ready for harvest. Moisture content of soybean at this stage is around 13 to 14%. The plants may be pulled by hand or cut with sickles and stacked in small heaps in the field to dry in the sun for few days. Then these are shifted to threshing floors. Threshing may be done either by beating the plants by sticks or with the help of suitable thresher with the caution that excessive beating should be avoided so that the seed is not injured. The seed is winnowed and cleaned after threshing and dried in to sun to 8 to 10% moisture content. Care should be taken to avoid mechanical admixture of seeds during these operations.

4.6.9 Seed yield

With good management practices seed yield of soybean varies from 20-25 q/ ha.

4.6.10 Storage

Soybean, to be used for seed in next season, should be stored under low temperature and low humidity conditions. The grains should be dried in sun and stored in ventilated rooms.

4.6.11 Quality

Soybean is an excellent protein source and called wonder bean. Composition of soybean grain is given in Table 4.2. It has been reported that black seeded soybean varieties are rich in protein but low in oil content, while yellow seeded varieties are rich in oil and poor in protein.

Table 4.2: Composition of soybean grain (values/100 g Edible portion)

Moisture	Protein	Fat	Minerals	Fibre	Carbohydrate	Energy	
8.1	43.2	19.5	4.6	3.7	20.9	432	
Ca	P	Iron	Carotene	Thiamine	Riboflavin	Niacin	Vitamin C
240	690	11.5	426	0.73	0.39	0.32	0.3

Important hints for obtaining maximum yields

(i) Use the improved variety recommended for the tract.

(ii) Use good and bold seeds for sowing.

(iii) Treat seed with fungicides.

(iv) Inoculate seed with a culture of *Rhizobium* in the soils for new introduction.

(v) Sow the seeds at the optimum time, using the optimum seed rate and optimum spacing.

(vi) Use adequate quantities of balanced fertilizers, depending upon the tract, variety, season etc.

(vii) Control weeds in time with hand weeding or by applying herbicides.

(viii) Adopt timely plant protection measures for controlling pests and diseases.

References

Joshi, O.P. and Billore, S.D. (1988). Herbicidal Weed Control to Soybean (*Glycine max*) Indian J. Agron. 43(1): 126-128.

Mondal, S.S. Sarkar, S and Pramanik, C.K.. (2004). Effect of K on Soil fertility and productivity under intercropped soybean (*Glycine max* L Merill) and sesame (*Sesamum indicum* L.). Tropical Agriculture 81(2): 66-71.

Mondal, S.S. and Pramanik, C.K. (1996). Integrated fertilizer management with potassium in soybean and sesame under different cropping system. Journal of Potassium Research 12: 298-304.

Rathore, R. (2002). Techniques and management of field crop production. Published by Agrobios (India): 308-316.

Rajput, R.L., Kaushik, J. P. and Verma, O.P. (1991). Yield and nutrient uptake in soybean (*Glycine max*) as affected by irrigation, phosphorus and row spacing.

4.7 Safflower

Safflower (*Carthamus tinctorius* Linn.) is an oilseed crop and popularly known as Kusum. Its oil consists of unsaturated fatty acids (94%). Safflower oil is good for heart patients due to its high percentage of linoleic acid (78%). The seeds are edible and are eaten after roasting. Being deep rooted and drought tolerant it is remunerative crop under dry land situation. The crop has also a good tolerance to the soil salinity as well. The oil is also used in paints and varnish industry. Their oil content varies from 24 to 36%, depending on the variety, soil, climate and other conditions.

4.7.1 Origin and distribution

According to Vavilov (1926), the centre of origin of safflower are India, Afganisthan and Ethiopia. *Carthamus oxycantha* is a wild safflower considered for its origin. The important safflower growing countries are India, USA, Mexico, Spain, the USSR and Australia. In India safflower is mainly grown between 14° and 45°N and 15° and 35°S. In India over 98% of the area is concentrated in the states of Maharashtra, Karnataka and Andhra Pradesh. It is also cultivated in Odisha and Madhya Pradesh in small scale. India ranks first with respect to area and second in production.

4.7.2 Botany

Safflower is thistle-like annual herb, highly branched, varying in height from 30 to 150 cm. It has many branches, each terminating in a flower. The inflorescence is broad, flat or slightly curved and densely bristled owing to the presence of numerous floral bracts. The flower's colour is yellow to red, containing 20 to 100 individual florets each of which produces a seed.

Classification

Safflower belongs to the family compositae and chromosome No. 2n=24. The related wild species are *Carthamus oxycantha* (wild safflower) and *Carthamus lanatus* (Thistle safflower). Based on flower's colour, it is classified into (i) white (ii) yellow, (iii) light orange and (iv) reddish orange.

4.7.3 Climate

Safflower is grown during *rabi* primarily as rainfed due to its deep root system. It is also grown as irrigated crop. Germination takes place at 15-16°C. For its high yield 25-30°C temperature during flowering is favourable for its high yield. Crop is tolerant to low temperature at seedling and vegetative phase. Low temperature at flowering cause an adverse effect on pollination and seed set. Crop is well grown in areas receiving 600-900 mm precipitation. High rainfall or prolonged rains at flowering reduce pollination, seed formation and impair quality of produce. Water-logging due to poor drainage or prolonged rains even for relatively short periods can cause substantial fall in seed yield.

4.7.4 Land requirements

Select lands on which same kind of crop was not sown in the previous season unless the crop of the previous season was of the same certified variety. The field should be deep, well aerated, and water retentive soils having neutral reaction. Being drought-resistant, it is cultivated on all types of soil, including sandy soil, but it thrives best on water retentive black soils and alluvial loams.

Land is prepared by two-three harrowing and planking to remove the weeds and break the clods. Seedbed must be well pulverized and free from stubbles. Mix well 20-25 kg/ha chlorpyriphos dust 4% at last ploughing to protect the crop from termites and other soil insects.

4.7.5 Isolation requirements

Isolation distance for foundation seed is 200 m and for certified seed it is 100 m from other fields growing the safflower crop.

4.7.6 Agronomic management

4.7.6.1 Sowing parameters

Rathore (1992) reported the crop is sown during *rabi* primarily as a rainfed crop, but in some areas it is grown under irrigation. Safflower is generally sown from last week of September to second fortnight of October in different parts of India. Early sowing are found beneficial particularly with regards to less infestation of aphids and other insects. Safflower is generally sown by using seed about 15-20 kg/ha. Row to row spacing is 45-60 cm depending upon climate and soil conditions. For heavy rainfall zone of Gujarat, Gujarat Agricultural University recommended sowing of 'Bhima' variety at 60 cm spacing using 15 kg seeds/ha. Plant to plant spacing may be kept at 30 cm and depth is 5-6 cm. When safflower grown in strips a row spacing of 45 cm is adapted.

Variety

Selection of variety according to soil and climatic conditions is important to harvest suitable yield. A list of improved varieties of safflower, suitable for different parts of the country is given below :

Table 4.3: Improved varieties of safflower cultivated in different regions of the country

Variety	Oil content (%)	Duration (Days)	Yield (q/ha)	Adaptability
Bhima	32.5	-	12-14	Maharastra
JFS 1	30.0	-	10-15	Rajasthan, Madhya Pradesh
CO 1	-	120-125	7-8	Tamil Nadu
N 62-8	30.0	140-150	10	Parts of Maharashtra
A 1	31.0	125-130	8-9	Andhra Pradesh and Karnataka
Manjira	32.5	105-110	8-10	Andhra Pradesh
T 65	30.0	-	12-13	Uttar Pradesh
K 1	-	120-125	6-8	Bihar, Uttar Pradesh and Tamil Nadu
Nira	30.0	-	14-18	Maharashtra and Tamil Nadu
HUS 305	35.0	-	11-13	Uttar Pradesh and Peninsular India
JSF 2	33.0	-	12-14	Madhya Pradesh
JL 28-1	31.4	114-167	18-20	Maharashtra
Tara	32.5	130	12-14	Maharashtra
A-300	31.9	125	7-8	Karnataka
Nag-7	30.0	140	10-12	Maharashtra

4.7.6.2 Nutrient management

For safflower cultivation, FYM @ 10-12 t/ha should be incorporation thoroughly 15-20 days prior to sowing in light soils. Crop responds well to applied nitrogen. The application of 20-40 kg N/ha results in a substantial increase in yield. Crop responds well to applied nitrogen. Response to phosphorus depends on inherent status and phosphorus fixing capacity of soil. Soils deficient in phosphorus, respond well to basal dressing of 20-25 kg P_2O_5/ha. Patel *et al.* (1994) obtained response of 25 kg P_2O_5/ha over control at Navsari, Gujarat. In areas with adequate moisture, 20 kg K_2O/ha may be applied. It has been reported that its requirement for N, P_2O_5 and K_2O/ha varies from 20-25, 15 to 50 and 20 to 40 kg/ha (Pasricha *et al.,* 1991).

4.7.6.3 Weed management

In general, the pure crop is given one or two weeding, combined with hoeing on the 20th and 45th days after sowing. Mishra (1997) reported 35-60% loss due to weeds in safflower. Weeds grow luxuriantly, because of wide row spacing and initial slow growth of the crop. Thus, 15-45 days of growth is considered as most critical from crop-weed interference view point. The most common weeds found in crop field are *Chenopodium album, Chenopodium murale, Anagallis arvensis, Spergula arvensis, Melilotus indica* as broad-leaf weeds and *Phalaris minor, Cynodon dactylon* as grassy weeds. Hand weeding at 20 and 35 days after sowing keep most of the weeds under control. Weeds could also be controlled by using herbicides in less time and economically. Singh and Singh (1996) reported pre-emergence application of pendimethalin @ 0.75 lit/ha equally effective to two hand weeding.

4.7.6.4 Irrigation management

Safflower is grown during *rabi* season primarily as a rainfed crop, but in some areas it is raised under irrigation. Rathore (1992) reported that safflower is deep rooted, xerophytic plant, and can survive under scarce soil moisture. Under stress conditions, one to two irrigations at 25 and 75 days after sowing enhances yield. Rosette and flowering are the most critical stages for irrigation. In water retentive soils zone, life-saving irrigation may yield good crop of safflower.

Patel and Patel (1993) reported 22, 51 and 68% increase in yield with one irrigation at grand growth at flowering and at seed formation stages respectively over no irrigation.

4.7.7 Nipping

When the plants have developed the central flowering head, they are topped to promote branching, flowering and seed production. Subbiah and Swaram (1965) recommended nipping of central shoot to induce branching and to promote seed yield.

4.7.8 Roguing

For seed production roguing must be done at least at vegetative, flowering and maturity stage. Before flowering rouging should be done on leaf and stem characters. After flowering rouging done on the basis of inflorescence or ear head characters and before harvesting rouging should be done on the basis of grain character.

4.7.9 Plant protection measures

The safflower aphids (*Uroleucon carthami*) infestation can be checked by early sowing. Spray the crop with dimethorate or monocrotophos (0.07%) for controlling aphids and bud fly. Rust (*Puccinia carthami*) and *Alternaria* leaf spot (*Alternaria carthami*) are the important diseases of safflower. The growing of resistant varieties is recommended to avoid damage from these diseases. Grow resistant varieties like 'APRR' and 'SF 438'. Seed treatment with mancozeb @ 3 g/kg of seed prevents infestation.

4.7.10 Harvesting and threshing

Safflower crop generally attend maturity in 4 or 5 months after sowing. Crop should be harvested when lower leaves turn yellow and start drying. The ripe plants are either cut-off from ground level with sickle or pulled-out and stacked for a few days to dry, then threshed by beating with sticks. The threshed material is winnowed to obtained clean seeds. The seed must be dried to obtain 92% moisture content for proper storage.

4.7.11 Yield

The seed yield under irrigated condition with improved technologies is 15-18 q/ha and in scanty moisture condition the yield is 8 to 12 q/ha. Under rainfed situation the seed yield is 6.7 q/ha.

Bird damage

Birds particularly parrots, pose a serious problem to safflower when it raised in isolated pockets as in non-traditional areas. To minimize damage from birds, take up safflower cultivation in large continguous blocks. Safeguard crop from bird damage during seed-filling to physiological maturity which extends by about three weeks through effective bird scaring, particularly in morning and evening hours.

Important hints for obtaining the maximum yields

(i) The sowing time should be adjusted in such a way that there will be no rainfall after flowering.

(ii) Nitrogen amounting to 40 kg/ha has to be applied preferably in two splits, half at the time of sowing and the remaining half at flowering.

(iii) Timely weeding for eradication of weeds and controls measure against pest and diseases should be adapted.

References

Patel, Z.G., Mehta, S.C. and Raj, V.C. (1994). Response of safflower (*Carthamus tinctorius*) to row spacing and nitrogen and phosphorus fertilizers in vertisols of South Gujarat. Indian J. Agron. 39(4) : 699-700.

Mishra, J. S. (1997). Herbicidal weed control in soybean. Indian Farming 47(3): 17-20.

Patel, Z.G. and Patel, N.C. (1993). Effect of irrigation scheduling on yield of safflower (*Carthamus tinctorius*). Indian Journal of Agronomy 38(3): 509-510.

Pasricha, N.S., Bahl, G.S., Aulakh, M.S. and Dhillon, K.S. (1991). In: Fertilizer-use Research in oilseed and Pulse crops in India. ICAR.

Singh, J. and Singh, B.S. (1996). Chemical, weed-less try in Safflower (*Carthamus tinctorius*). Indian Journal of Agronomy 40(4): 510-514.

Subbiah, N. and Swaram, M.K. (1965). Increasing yield in safflower. Indian Oilseed J. 9:155.

Mishra, J.S. (1997). Critical period of weed competition and losses due to weeds in major field crops. Farmer and Parliament. 33(6):19-20.

Vavilov, N.I. (1926). Studies on the origin of cultivated plants. Bulletin of applied Botany and Plant Breeding. 16(2): 61-64.

Rathore, R.(2002). Techniques and managemcnt field crop production. Publishing by Agrobios (India).

4.8 Castor

Castor is one of the most commercial non-edible oilseed crops in the country. The oil content varies from 35-58%. Castor oil is being used widely for various purposes. It is used as a lubricant in high-speed engines and aeroplanes, in the manufacturer of soaps, transparent paper, printing-inks, varnishes. It is also used for medicinal and lighting purposes. The oil cake is used as manure and plant stacks as fuel or for preparation of paper pulp. The castor leaves are used for rearing silkworms in silk producing areas. Castor oil is used in the manufacture resins.

4.8.1 Origin and distribution

According to available literature, castor (*Ricinus communis L.)* is indigenous to Eastern Africa and most properly originated in Ethiopia (Weiss, 1971). Russian plant breeders considered that there are four large centres for variability, each with its own specific plant characters to support a polyphyletic origin for the crop. The most important is the Irano-Afganistan-USSR region. The other three located in Palestine, South-West Asia, India-China and Arabian Peninsula (Moshkin, 1977). The main castor growing countries are Brazil, India, the USSR and Argentina. In India major castor growing states in India are Gujarat, Andhra Pradesh, Maharastra, Karnataka and Madhya Pradesh and Bihar.

4.8.2 Botany

Castor belongs to the genus *Ricinus*, a member of Euphorbiaceae which contains a vast number of plants mostly native to the tropics. The genus *Ricinus* is considered to be monotypic and *R. communis* is the only species which includes many polymorphic types. The cultivated types are dwarf annuals. The stem is erect, circular in section, partially hollow, smooth, glabrous, with good branching. The lower nodes are shorter and their length increases with the height. Leaves are alternate, large, palmate with 5-11 lobes, acuminate, margin notched, serrate or indented. They are carried on long stout petioles. The inflorescences are borne terminally on the main and lateral branches. Flowers are large, in terminal sub-panicled recems, monoecious apetalous, the upper portion of the receme being occupied by the female flowers and the lower by male flowers. The fruit is a roundish glaucous capsule, with three projecting sides covered with tough spines or smooth, three loculated and three seeded. Seeds are albuminous, anatropous, broad and oval. The testa is brittle, varying in colour and mottling. Below the testa, there is the thin stamen, covering the whitish oily endosperm containing the embryo.

Classification

Popova (1930), a Russian worker, tried to divide the species *communis* into the following sub-species. This classification is based on eco-geographical groupings. But the last three resembles each other that it is very difficult to differentiate. Hence, it is suggested that the last three be placed under one group namely Indo-African.

(i) **The Persian sub-species : Percius G. Popova:** It is found in Iran, India, Middle Asia and parts of Russia. This sub-species has two outstanding morphological characteristics. The inflorescence is elongated, narrow, conical and the fruits may cover the flower stalk to the base.

(ii) **The Chinese sub-species : Chinensis G. Popova:** The forms of Chinese sub-species are found in West China and Manchuria. This sub-species is early maturing and takes 75 to 97 days for maturation. The inflorescence is short and loose with a broad, conical shape. Seed is oval and large, and grey coloured with a small caruncle.

(iii) **The Indo-African sub-species : Africanus G. Popova:** The sub-species are found in four continents. The ecological diversity of this sub-species is very high. It possesses 88% of all the known characters of castor plant and there it is very valuable for breeding purposes.

(iv) **The Zanzibar sub-species : Zanzibarinus G. Popova:** The sub-species are found in the tropical parts of Africa and America and in Europe as ornamental plants. Most form of this subspecies is late maturing. The inflorescence is cylindrical or ellipsoid in shape.

4.8.3 Climate

Castor is tolerant to drought and grows well in relatively dry, warm regions, having a well distributed rainfall of 50 to 75 cm. In heavy rainfall areas, the crop puts on excessive vegetative growth and assumes a perennial habit. The crop cannot stand frost, but some annual varieties can grow even at altitudes of 1200 to 2100 m. Castor requires a moderately high temperature 20-26°C with low humidity throughout the growing season to produce maximum yields. It grows in tropical and subtropical regions as a perennial plant and in temperate climate as an annual plant. Long, clear sunny days are most suitable and cloudy or humid days irrespective of temperature will reduce yields.

4.8.4 Land requirements

The land should be selected for seed production of castor, on which castor was not grown in previous season, unless the crop grown in previous was of the same variety and grown from certified seed. In addition field should be well drained and well-aerated.

4.8.5 Isolation requirements

Castor is an often cross-pollinated crop. Cross-pollination by wind varies from 5 to 36% according to prevailing climatic conditions. The seed field must be isolated from other fields of castor at least by 300 meters for foundation seed class and 150 meters for certified seed class.

4.8.6 Agronomic management

4.8.6.1 Soil and land preparation

The crop can be grown well on almost all types of soils, but it is generally grown on red sandy loams in peninsular India and on light alluvial soils in the Northern states. Deep moderately fertile with slight acidic conditions are more suitable. The castor grows well in soils, which is not fit for valuable commercial food crops. Deep ploughing is beneficial for rainfed castor because it helps in absorption and conservation of rain water. Harrowing once or twice with the receipt of pre-monsoon rains is advantageous. So, in general one deep ploughing followed by two to three harrowing with disc or blade harrows is sufficient to bring the field to the desired tilth for cultivation.

4.8.6.2 Sowing parameters

The rainfed castor is sown with the onset of monsoon. The sowing time of castor in most of the states is June to July. In Bihar, castor as *rabi* crop is sown in September to October. In Gujarat, planting is done in August to September. In Karnataka, castor is sown in April.

Cultivation should included nucleus / breeder / foundation seeds for planting from source approved by seed certification agency. Castor is sown in rows either by drill or behind the plough in furrows and it can be grown in dibbling method. Dibbling method is particularly preferred for seed production and hybrid cultivation. The depth of seeding is 8 to 10 cm for rainfed crop and 6 to 7 cm for irrigated crop. For line sowing, row to row spacing is 90 cm and plant to plant spacing is 45 to 60 cm.

The rate varies upon spacing, seed size and methods of sowing. Generally, a seed rate of 12-18 kg/ha is sufficient. For rainfed crops higher seed rate is needed, about 33-44 kg/ha is practiced. Before sowing, castor seed should be treated with Mancozeb @ 3 g/kg of seed to prevent the attack of root rot and *Alternaria* blight.

4.8.6.3 Nutrient management

The castor crop responds well to organic manure. Application of FYM or compost at the rate of 5-6 t/ha during the last ploughing is beneficial to improve the soil properties and it also help to enhance the availability of micro-nutrients. The rate of N, P and K application varies considerably in different states and ranges from 40-80 kg N, 40-50 kg P_2O_5 and 20-40 kg K_2O/ha. The entire dose of P and K and half of the nitrogen is applied as basal and remaining half of the nitrogen about 30 days after sowing for better seed production.

4.8.6.4 Irrigation management

Castor is generally grown in *kharif* season as rainfed crop. If the onset of monsoon, amount and distribution of rains during the growing period is normal, the crop growth and yields would be upto the mark.

4.8.6.5 Weed management

Castor yields are highly affected by the weed competitions in early stage of growth i.e. upto 45 to 60 days after sowing. Different types of weed flora are seen in castor field i.e. *Brachiaria* spp. *Echinochloa crusgalli, Digera arvensis, Argemone maxicana, Leucas aspera, Cynodon dactylon* and *Cyperus rotundus*. The castor field must be kept weed free for the first 60 days after planting. Two to three weedings/hoeing are sufficient to keep the field clean. In castor field weeds can also be controlled by using the following herbicides.

Soil application of fluchloralin @ 1 kg/ha as pre-sowing or pendimethalin 1.5 kg/ha as pre-emergence to crop and weed (1-2 days after sowing) is found to be beneficial.

4.8.7 Variety

The improved varieties of castor recommended for cultivation in different states are given below.

Table 4.4: Improved varieties of castor recommended for different states

Variety	Characteristics	State
Aruna	Dwarf yields 1300 kg/ha with 50.9% oil; duration is 150-160 days	Andhra Pradesh, Karnataka, Odisha, Rajasthan
GAUCH-1	Moderate branching, triple bloom; duration is 150 days, yield 1500 kg/ha, oil content 49.5 %	Gujarat
GCH-3	Duration 150 days, Dwarf, heavy branching; yield 1500-1600 kg/ha. Oil content 49.5 %	Gujarat
Bhagya	Duration 120-150 days, Dwarf, large seeded yield 2000 kg/ha, oil content 53%	Andhra Pradesh
Saubhagya	Duration 240-270 days, Dwarf type with condensed internodes, yield 2500 kg/ha; oil content 50 %	Andhra Pradesh
Punjab Castor No.1	Duration 275 days, Tall, yields 1700 kg/ha; oil content 52%	Haryana
TMV-1	Duration 195 days, yields 900-1000 kg/ha; oil content 51%	Tamil Nadu
Tarai-4	Duration 240 days, Tall, yields 1100 kg/ha; oil content 52%	Uttar Pradesh
B-1	Duration 130 days, Dwarf, non-spiny capsules; yield 300 kg/ha; oil content 43.5 %	West Bengal
EB. 16A	Duration 220 days, tall, sparsely branched, yield 1450 kg/ha, oil content 50.9%	Bihar

Irrigations requirement varies with the rain. Usually two to three irrigations during the entire crop season may be required. Adequate soil moisture at the time of flowering is necessary for better production of castor seed.

4.8.8 Plant protection measures

Insect-pest

Castor semi-looper (*Achoea janata* Linn) – The castor semi-looper fed voraciously on castor leaves. This damage can be controlled by spraying Monocrotophos 36 SL @ 1 ml per litre of water or Quinalphos 25 EC @ 2.0 ml per litre of water.

Castor capsule borer (*Dichocrocis punctiferalis* Guenec) – This capsule borer bores in the capsules and eats the seed causing excessive damage to the crop. This damage can be controlled by spraying Monocrotophos 36 SL @ 1.0 ml per litre of water.

Castor hairy caterpillar (*Euproctis lunata* Walker) – Caterpillars feed on leaves of plants and in case of severe infestation, they may cause complete defoliation. It can be controlled by dusting the crop with Quinalphos 1.5% dust.

Diseases

Seedling blight (*Phytophthora parasitica* Dustur) – The first indication of the disease is the appearance of a roundest patch of unhealthy dull green colour on both the surfaces of cotyledons of seedling of 15-20 cm high. It can be controlled by seed treatment with mancozeb @ 3 g/kg of seed.

4.8.9 Roguing

All the off-type plants should be removed from the field before flowering for pure seed production. The diseased plants affected by *Phytophthora* blight and *Cercospora* leaf spot should be rouged out as soon they are noticed. Preventive plant protection measures adapted immediately to check further spread.

4.8.10 Harvesting and threshing

The fruits of castor are ready for picking in November. The picking is done continues until April, because the capsules mature unevenly due to sequential development of racemes. The fruit should be gathered when they start turning light yellow and placed in piles to dry in the sun until they blacken. Later the seeds are beaten out with sticks, winnowed and screened to remove husks, dry skins and soil pebbles. Before storage the seed must be dried to at least 8% moisture content.

4.8.11 Seed yield

Average yield is 8-10 q/ha. High yielding varieties with good management practices give 15 to 20 q/ha seed yield under irrigated condition.

Important hints for getting maximum yields

(i) Recommended variety for a particular area should be chosen.

(ii) Seed treatment with fungicide should be done before sowing to prevent the attack of root rot and alternaria blight.

(iii) Sowing should be done at the time recommended for different agro-climatic conditions. Sowing should be done with proper spacing recommended for different regions.

(iv) The recommended fertilizer schedule for the region should be adapted.

(v) Adequate plant protection measures should be adapted.

(vi) Two to three inter-cultivation should be done to provide aeration for the root-system and to keep the crop free from weeds.

(vii) The nipping of branches increases the size of the main spike and induces early maturity.

(viii) Harvesting must be done at the right time, as too early or too late harvesting results in reduced yield, oil and shattering of seed respectively.

References

Moshkin, V.A. (1977). On the origin and distribution of castor as a crop. Byull 69, Inst. Rast. N.I. Vavilova, Rosyov-na-Don, USSR.

Popova, G.M. (1930). Castor oil plant. Lenin Acad. Agril. Sci. Inst. Appl. Bot. and New Culture Leningrad. p. 63.

Weiss, E.A. (1971). Castor, sesame and safflower. Leonard Hill, London : 1-901.

5

Seed Production of Fiber Crops

5.1 Jute Seed Production

Jute ranks secand next to cotton as the most important natural fibre and it recognized as the 'golden fibre'. Its products being biodegradable and environment friendly in contrast to synthetic substitutes, a renewed interest in this natural fibre has been generated in the recent past throughout the world. Important characteristics of jute fibre are its silky lustures, high tensile strength, low extensibility, considerable heat resistance. Because of its great abundance and shortage of cotton, jute is blended with cotton and jute/cotton blended yarns are produced mainly to take advantage of the lower price of jute fibre (Doraiswamy *et al.*, 1998). Several diversified product is also made from jute. Since currently the paper industry has been using it as a raw material for paper. It is also being used in geotextile. The jute sticks are largely used as fuel and also for making gun powder.

At present the jute acreage in India has stabilized around 8 lakh hectares with a total production of 100 lakh bales and the national productivity has stepped up to 22 q/ha. Jute being a labour intensive crop, about 40 lakh farm families derive their subsistence by cultivating jute in the country. Moreover, about 2.5 lakh people are employed in the jute industry and 25 lakhs people are engaged in jute based ancillary sector. India earns 500-600 crore rupees as foreign exchange by exporting jute. So, cultivation of jute shares a major role in Indian economy.

5.1.1 Origin and Distribution

The primary centre of origin of *Corchorus olitorius* is considered to be in Africa, while secondary centre may be in India or within the Indo-Burma region. The centre of origin of *Corchorus capsularis* is in the Indo-Burma region (Kundu, 1951). It has been successfully grown in India, Bangladesh, Nepal, China, Indonesia, Thailand, Mynamar (Burma) and some South American countries. The major jute growing states in India are West Bengal, Bihar, Assam, Odisha, Uttar Pradesh, Meghalaya and Tripura.

5.1.2 Botany

The genus *Corchurus* of the Tiliaceae family includes about 40 species distributed mostly in tropical region. Out of that, the species *C. capsularis* and *C. olitorius* are cultivated. Both species are herbaceous annuals. Depending on the time of sowing, variety, soil and climatic conditions, the vegetative growth period of both is 3-5 months; at the harvesting stage, varieties of the *C. capsularis* attain a height of 5-12 ft and *C. olitorius* 5-15 ft or more. It has tap root with side root profusely developed in upper soil layer. The stem of both are cylindrical. The initiation of the reproduction phase is marked by branching of the apex in contrast to the production of side branches during the vegetative phase. Leaves of *C. olitorius* are glabrous, narrower and usually lack the glucosides responsible for bitter taste of *C. capsularis* leaves (Anon, 1956). The flowers of *C. capsularis* are smaller and open to two hours after Sunrise, while those of *C. olitorius* are large and bloom before Sunrise. Flowers of both species are yellow or pale yellow. Inflorescence is extra-axillary, usually opposite leaf base, cymose, which normally has three flowers. Two to three of the flowers develop into seeded fruit in *C. capsularis*, while usually one, sometime two or occassionally three develop into fruit in *C. olitorius*. The fruit is globular with 5 locular containing 35-50 chocolate brown colour seeds in *C. capsularis* and cylindrical with 5-6 locular containing 140-240 bluish green to steel grey or black colour seeds in *C. olitorius*. Fruit shape is main diagnostic difference between the two species. Spotaneous morphological variation in *C. capsularis* is greater than that in *C. olitorius* (Chaudhuri, 1933; Kundu and Sharma, 1956).

Both jute species are short day plants with a critical photoperiod of about 12.5 h. Flowering is delayed by exposure to longer photoperiods and photoperiods of 10-12 h induce flowering in 30-35 days. Very early sowing causes premature flowering in both species because of the prevailing short day, this is followed by a second vegetative period and by flowering at the normal time (Mukherjee, 1948). *C. olitorius* is more sensitive to short days than *C. capsularis*. In *C. capsularis* the narrow range of flowering period continued to about 4 weeks with maximum flush of flowering in the second week, whereas in *C. olitorius* range of flowering period is prolonged as well as the flowering phase and vegetative phase co-exist during 8-10 weeks and flowering occurs in periodic flushes (Kar, 1959).

Initial studies on the reasons for incompatibility between *C. capsularis* and *C. olitorius* showed that pollen tubes in reciprocal crosses grew through the styles normally (Srinath and Kundu, 1952), fertilization took place and a few celled embryo and some free endosperm nuclei were formed; however, the embryo degenerated early in development so that seed failed to set (Ganesan *et al.*, 1957; Patel and Datta, 1960). It has recently been stated that in crosses involving grafts of the two species, the percentage of fruit set was similar, irrespective of whether *C. capsularis* or *C. olitorius* was the scion (and the pistillate parent) provided the pollect parent was non-bitter, i.e. a non-bitter strain of capsularis or strains of *C. olitorius*, almost all of which are non-bitter. In other cases the cross failed when *C. olitorius* was the pollen parent. As graft survival was also better when *C. olitorius* was the scion, it was concluded that the two species differ in their bio-chemical constitution and that sexual and vegetative incompability may go together (Bhaduri and Bairagi, 1969).

Natural crossing in jute according to earlier worker such as Finlow and Burkill (1912) and Howard *et al.* (1919) is a rare phenomenon and self-fertilization is the rule in both the cultivated species of jute – olitorius and capsularis (Finlow *et al.*, 1919). Dutta and Ghose (1962) and Basak and Chaudhuri (1966) showed that considerable extent of natural crossing takes place in jute. Percentage of natural crossing in olitorius jute vary in different years as well as among the plants. It had been reported an average 6.1 – 12.8% natural crossing in olitorius jute and 1.4-8.9% in capsularis jute (Ghosh and Dasgupta, 1945).

From an experiment Basak and Gupta (1972) reported that the mating system of the olitorius jute was partial inbreeding i.e. mixed system of random mating and self fertilization. It was also found that natural crossing in both the two species was affected through the agency of insects, and climatic factors particularly rain during the flowering period.

Hybridization of the two species at the tetraploid level or in diploid or tetraploid combinations has not been very successful and there is little evidence that an increase in the number of genomes confers much advantage to the breeder or reduces to any great extent the incompatibility between the species (Rao *et al.,* 1944; Bhadurai and Chakraborty, 1948; Nakajima, 1949; Chaudhuri *et al.*, 1956; Kundu and Rao, 1956; Datta, 1963).

Banerjee and Rakshit (1970) reported that the male-sterile 'Ribbon' is effective in inducing heterosis in jute. The production of hybrid seeds of jute may be significantly increased by decreasing the distance between the seed and the pollinator parents and by increasing the number of seed presents. The number of seed parents can be increased if seeds from a cross between 'Ribbon' and its heterozygous normal sibs are used, when a segregation of 1 'Ribbon' and 1 normal is expected. An increased seed rate may also give more 'Ribbon' plants.

5.1.3 Climatic Requirement

For production of seed jute is grown in rainy season when the temperature is 21-28°C and the relative humidity (RH) is 60-95%. It grows in areas receiving an annual rainfall of 100-250 cm, about 150 cm being optimum, about 40-75 cm of rain should fall intermittently during the first 6 weeks from land preparation.

Jute seed production can be grown in areas suited to fibre production. But in order to avoid seed-fibre competition, it is better to cultivate beyond the fibre production zone. High humidity may favour growth and development; but at harvest; high humidity may cause fungal infection of seed and thus deteriorating the seed viability as well as its quality (JARI, 1973).

5.1.4 Soil Requirement

The land for seed production should be medium to high uniform, well leveled, fertile, slightly acidic to neutral (pH 6-7.5) sandy loam to loam soil with good drainage. The land should be free from obnoxious weeds and soil borne diseases (root rot and stem rot). The field selected for certified seed production should be one on which same type of crop of other variety was not grown in previous season. If same variety of certified seed jute or other kind of crop was grown in previous season may be selected for certified seed production (Agarwal, 1980).

5.1.5 Isolation

Jute is normally a self fertilized crop, although in *C. capsularis*, the percentage of natural crossing is very low (not more than 25%) and in *C. olitorius*, it is however, observed to the extent of 12% (Ghose and Dasgupta, 1945; Roy, 1960 and Ghosh and Sen, 1974). This amount of natural crossing in *C. olitorius* creates a problem in maintaining genetic purity in seed production programme. The results in experimentation for determination of isolation distance between two varieties of *C. olitorius* to avoid natural cross pollination indicated of a distance of 7.5 metres from the source of contamination, the natural crossing was negligible (0.01 to 9%). Though natural cross pollination in *C. capsularis* is very low it will be safe to maintain an isolation distance of 10 metres between varieties in both this species of jute.

The natural cross pollination is likely to generates genetic variability in the population, hence, a knowledge of the extent and made of cross pollination is essential for improving the existing breeding method and also for maintenance of breeder's stock and foundation stock of released varieties and germplasm.

For jute seed production of both species, the seed field should be isolated by a distance of fifty meters for foundation seed and thirty meters for certified seed from the jute field of other varieties of same species as well as from the field of same variety which is not conforming to genetic purity of variety for certification of seed.

Plant breeders aim towards development of crop varieties with potentiality for high yield of economic product and desired attributes like adaptability, resistance/tolerance to biotic or abiotic stress etc. Jute germplasm have been effectively utilized for development of improved cultivars (Table 5.1 and 5.2).

5.1.6 Agronomic Management

In case of jute, improved seed alone was found to increase the fibre yield by more than 16% over that available from the use of the local seeds. However, the jute seed production is not so simple job as it seems to be. Unlike cereals, seeds and fibre both cannot be economically produced from the same jute plant, whereas in cereals, the end product may be used as seed and the package of practices for crop production and seed production do not differ. There certainly exists some difference in the package of practices in jute seed production and fibre production.

5.1.6.1 Land preparation

The land should be prepared well in order to get fine tilth. Usually one deep ploughing and three to four cross harrowing each followed by laddering will be enough, when soil moisture is optimum. Weeds and stubbles of the previous crop must be removed during land preparation to get a fine tilth. Thoroughly prepare soils of fine tilth is necessary to ensure good germination and less infestation of weeds (Saraswat and Pal, 1972; Agarwal, 1980).

5.1.6.2 Sowing parameter

For raising seed crop, primary requirement is of good quality seed (Genetically pure seed). For raising foundation and certified seed crop obtain nucleous or breeder and foundation seed respectively from an approved sources. Hence, the seeds should always be obtained from authentic sources like CRIJAF (Central Research Institute for Jute and Allied Fibre), SSC (State Seed Corporation), National Seed Corporation (NSC) etc. The germination percentage of the seed should be tested in petridishes before sowing in the field. This should be atleast 90 per cent. A good sample should have 98 per cent germination at 33°C temperature. Only the standard varieties should be used for raising seed crop. *C. capsularis* jute seeds take about 24 hours and *C. olitorius* jute seeds take about 48 hours for germination in petridishes (Sarswat and Pal, 1972).

Table 5.1: Released varieties of tossa jute (*C. olitorius*) and their pedigree

Name of variety	Popular Name	Pedigree	Year of release
JRO 632	Baisakhi Tossa	Selection from indigenous type	1954
JRO 878	Chaitali Tossa	JRO 620 x Sudan green	1967
JRO 7835	Basudev	JRO 620 x Sudan green	1971
JRO 524	Navin	Sudan green x JRO 632	1977
TJ 40	Mahadev	Selection from inter-mutant cross	1983
JRO 3690	Savitri	Tobacco leaf x Short internode	1985
KOM 62	Rebati	Gamma-ray derivative of JRO 878	1992
JRO 66	Golden Jubilee Tossa	Selection from multiple cross involving CG, TM, JRO 524, Peking, Bangkok and Tanganyika 1 exotic parents	1977
JRO 8432	Shakti Tossa	IC 15901 x Tanganyika 1	1999
JRO 128	Surya Tossa	TJ 6 x Tanganyika-1	2002
	Bidhan Rupali	Selection from yellow mutant, x-ray derivative of JRO 632	2002
S 19	Subala	JRO 620 x Sudan green x Tanganyika-1	2004

Source: CRIJAF, Barackpore, West Bengal.

Table 5.2: Released varieties of white jute (*C. capsularis*) and their pedigree

Name of variety	Popular Name	Pedigree	Year of release
JRC 321	Sonali	Selection from Hewti	1954
JRC 212	Sabujsona	Selection from an indigenous type	1954
JRC 7447	Shyamali	X-ray derivative of JRC 212	1971
JRC 4444	Baldev	JRC 212 x D 154	1980
UPC 94	Reshma	JRC 321 x JRC 212	1983
Hybrid C	Padma	JRC 6165 x JRC 412	1983
KC 1	Jaydev	Gamma-ray derivative sof JRC 4444	1992
KTC 1	Rajendra	Selection from an indigenous germplasm IC 30740	1994
JRC 698	Shrabanti	Selection from a multiple cross involving JRC 321, Tripura cap, JRC 212, JRC 918, Cordate leaf, Chinese 1, Egypt S5, EC 41336, EC 101572, Cap. Main land China, Japanese green, Cap. Non-branching and EC 117161.	1999
Bidhan Pat 3		D 154 x D 18 (photo insensitive mutant)	2000
Bidhan Pat 1		Gamma-ray derivative of D 154	2001
Bidhan Pat 2		D 154 x D 18 (photo insensitive mutant)	2001
JRC 80	Mitali	CIM 114 x JRC 321	2004

Source: CRIJAF, Barackpore, West Bengal.

Recently released varieties / hybrids of jute

Variety	Duration (days)	Yield (t/ha)	Area of adoption	Salient features
BCCC-2 (Bidhan Pat 5)	120 (fibre); 170–175 (seed production)	2.95 (16.4 bales/ha)	West Bengal	Suitable for rainfed, mid and high land areas followed by transplanted paddy, tolerant to stem rot disease, semilooper, BHC and Apion.
Central Kenaf (JRHC-3)	160–175	2.86	West Bengal, Odisha, Bihar, Andhra Pradesh, Maharashtra and North Eastern States especially Tripura and Assam	Suitable for rainfed cultivation, tolerant to foot and stem rot and YVMV.
JROMU-1	140–150	3.29	West Bengal, Assam, Bihar, Odisha	Suitable for jute growing belt for cultivation, tolerant to stem rot, Apion, semi-looper, BHC and yellow mite.

The sowing may be done in May for optimum seed production of jute. However, pruning is desired if early sowing is done (JARI, 1973). Time of planting significantly influenced the seed yield and yield attributing characters like the number of pods per plant, number of seeds per pod in both the species of jute (Mian and Gain, 1971). Early sowing of jute crop gives maximum vegetative growth whereas seed crop favours stunted growth with maximum lateral branches. So, early sown crop do not yield sufficient seed. Late sown crop in May-June give a higher seed yield (Kundu *et al.*, 1959; Nath, 1980). Das *et al.* (1959) observed that 2nd June sown crop (var. JRO-632) gave the highest seed yield closely followed by 2nd July, 17th June and 18th May. It was also reported that sowing of jute seed crops can be conventionally delayed upto the beginning of July. Singh *et al.* (1984) reported that the *C. olitorius* variety JRO-7835 gave the highest seed yield (8-10 q/ha) in both 30th June and 30th July sowing. However, JRO-632 gave lowest seed yield in both dates. Whereas, in *C. capsularis* variety JRC 7447 gave highest seed yield of 6.56 q/ha followed by JRC 321 and JRC 212 in June sowing crop. Joseph *et al.* (1970) reported that the seed yield of JRO 524 decreased due to delay in sowing beyond 3rd week of May. Thakuria and Sharma (1991) reported that Jute cv. JRC 212, sown on eight dates of 15 days interval between 1st June and 13th September gave the highest two years average (1986-87) seed yield of 11.6-11.8 q/ha when sown on 1st two dates. Further delay in sowing produced linear decrease in yield.

The experimental results obtained at Jute Agricultural Research Institute, indicated specific difference among *C. capsularis* and *C. olitorius* varieties in response to the natural sequence of day length and temperature. In January, February, March sowings with low temperature and short day lengths (10 to 12 hrs.) *C. olitorius* varieties showed early flowering within 45 to 60 days. While in *C. capsularis* varieties, the flowering was very much delayed widely varying from 110 to 180 days. In April, May and June sowings, with increasing day lengths above 12 to 14 hrs and high temperature, there was a progressive increase in growth, height and the initiation of flowering was also delayed, ranging nearly from 80 days to about 122 days. During this period both *C. olitorius* and *C. capsularis* varieties behaved more or less similarly. In July, August and September sowings the flowering time was enhanced in both *C. capsularis* and *C. olitorius* varieties with the progressive shorting of vegetative period. In October, November and December sowings, the *C. capsularis* and *C. olitorius* varieties behaved similarly as in January, February and March sowings. All the varieties of both *C. capsularis* and *C. olitorius* responded to short day periods, showing earlyness in flowering. Under long day condition, when plants were subjected to more than 12 hrs light period, the initiation of flowering was delayed (JARI, 1958-1959 and 1959-1960).

Mian and Gain (1971) studied the effect of time of sowing on the yield of jute seed. Both the cultivated species of jute (*C. capsularis* and *C. olitorius*) were sown of monthly intervals from May to December. Seed yield of both the species were highest in crops sown in Jute followed by those in May. The number of fruits per plant and seed yields decreased between June and September after which they increased. From the experiment it was suggested that jute crops requires long days for vegetative growth and short days for flower initiation and seed formation.

Joseph *et al.* (1970) studied the effect of sowing time on seed yield in 'Tossa' jute (*C. olitorius*). They found that sowing in late June, JRO-5408 gave the highest average seed yield (0.67 t/ha).

The *C. capsularis* varieties JRC 212 and JRC 321 gave their highest yield when sown on 25th April but *C. olitorius* varieties JRO 632 and JRO 7835 gave their highest yields when sown on 10th May. JRO 632 and JRO 7835 gave lower yields than JRC 212 and JRC 321.

Singh *et al.* (1984) reported that seed yields were highest in Jute cv. JRC 212 when sown on 15th May and in cv. JRO 7835 when sown between 15th June and 15th July.

Jute can be sown either by broadcasting method or by drill sowing method. Drill sowing is preferred. Line sowing of both species of jute facilities intercultural operations, field inspection and voguing.

In broadcasting method, half of the seed is sown along the field and the rest half across the field to ensure uniform distribution. A light laddering is necessary to cover the soil at uniform depth. As the jute seed is very small, so it should be sown 3-4 cm deep.

Seed crop favours stunted growth with maximum lateral branches, wider spacing helps to produce more branches per plant which increase the seed yield. To harvest a good seed crop spacing 30 cm in between rows and 15-20 cm within row should be maintained (JARI, 1973, Nath, 1980). Broadcasted crop should be thinned properly to maintain spacing 12-15 cm from plant to plant.

Seed treatment

Before sowing jute seed should be treated with fungicides against seed borne pathogen @ 5 g Dithane M-45 or Bavistin @ 2 g per kg of seed. The seeds should be thoroughly mixed in a rotary drum for ten minutes.

5.1.6.3 Roguing

For seed production roguing is must to maintain the genetic purity of seed. Off types, diseased and pests infected plants in the jute field should be uprooted in three different growth stages of the crop. The first roguing should be done at 35 to 45 days after sowing. Then second roguing should be done at flowering stage and the third must be done at capsule maturing stage.

5.1.6.4 Weeding and thinning

Being a monsoon crop, weeds pose a serious problem in jute. The most important interculture operation of jute includes timely removal of weeds and thinning. The following weeds were found prevalent in jute fields: *Echinochloa colonum*, *Leptochola filiformis, Echinochloa crusgalli*, *Eleusine indica*, *Digitaria sanguinalis*, *Cyperus rotundus*, *Boerhaavia diffusa*, *Physalis minima*, *Euphorbia hirta*, *Phylanthus niruri*, *Amaranthus viridis* and *Amaranthus spinosus*. In general, grasses and sedges constitute around 60 and 35 per cent of the weed flora respectively. The first weeding and thinning should be completed between 15-21 DAS. Second weeding and final thinning must be carried out by 15-20 days after 1st one followed by light hoeing for improving tilth, aeration and to preserve moisture. Since, the manual weeding is costly and time consuming, chemical weeding may be done with fluchloralin (Basalin) @ 1 to 1.5 litre/ha in sandy soils and 1.5 to 2.5 litre/ha in heavy soils on one day before or one day after sowing on moist soil. Fluchloralin is very much effective in controlling annual granes. In this endeavour, one post emergence herbicide (Quizalofop ethyl) and one pre-planting herbicide (Trifluralin) available commercially were found to be beneficial against the predominant weeds of jute. Application of quizalafop ethyl 5% EC (Targasuper or Whipsuper) @ 2.0 ml/lit at 15 to 21 days after emergence was effective against grassy weeds throughout the crop growth. Preplanting soil incorporation of Trifluralin @ 1.0 kg a.i./ha at one or two days before sowing was found to be effective in controlling both grasses and broadleaved weeds until harvest.

5.1.6.5 Nutrient management

Fertilizer increased yield of all *C. capsularis* varieties more than *C. olitorius* varieties, but the interaction between variety and fertilizer was not significant (Das and Kundu, 1966).

For seed crops of jute, the nitrogen used should be judiciously reduced to 15 kg/ha. It is advantageous to have short plants with more branches. Seed crop requires a basal application of potassium and phosphorus @ 15 kg/ha to produce healthy seed. Researchers studied the effect of time and methods of nitrogen application on the seed yield of jute in Terai humid region and observed that application of 30 kg N/ha at sowing and 30 kg N/ha to soil 45 DAS, gave the highest average seed yield of 800 kg/ha compared with 570 kg/ha without N. Application of N in split dressing at other times and also as a foliar spray gave lower yield.

Gupta (1959) studied the effects of various doses of nitrogen or jute seed yield in Tripura and observed that application of 80 kg N/ha increased the seed yields from 159 to 263 kg/ha in *C. capsularis* (cv JRC-7447) and 126 to 205 kg/ha in *C. olitorius* (cv. JRO-534); further addition of nitrogen over 80 kg was not significant.

Hossain *et al.* (1975) studied the effect of nitrogen and phosphorus on seed yield of jute. Average seed yields were 5.01 and 8.07 pounds per acre for *C. capsularis* and *C. olitorius* respectively. N + P gave higher yield increments than N or P alone. Optimum seed yields were achieved with 25 or 50 lb N + 50 lb P_2O_5/acre.

Increased seed yield appeared to be more associated with increased 1000 seed weight than with more pods per plant and seeds per pod. Among the major elements N is the most important. With N deficiency, growth is poor, the leaves are pale green, distorted and drop prematurely; also the number of internodes and the number and size of flowers and fruits are reduced. In trials with varieties of *C. capsularis* and *C. olitorius* during three years, application of 33.6 kg N, 44.8 kg P and 44.8 kg K/ha increased the seed yield of *C. capsularis* by 98% and *C. olitorius* by 70% (Das and Kundu, 1966).

Rates of N upto 67.2 kg/ha were beneficial to *C. capsularis* but higher rates were uneconomic and sometimes even reduced yield (Kundu, 1956).

Ray and Majumdar (1995) reported that nitrogen level influenced seed yield of white jute (*Corchorus capsularis*) and tossa jute (*C. olitorius*). Five *C. capsularis* and five *C. olitorius* cultivars were sown at Mohanpur, West Bengal and gave highest seed yield in *C. capsularis* at 60 kg N/ha; while seed yield decreased at highest N rate. In *C. olitorius* there was no significant difference between N rates of 40 and 80 kg/ha.

Jute for seed production requires a basal dressing of phosphorus @ 15 kg/ha to produce healthy seed (JARI, 1973). *C. capsularis* has been reported to show a greater response to P than *C. olitorius* (Gupta and Sen, 1953). With P deficiency, plants are stunted, the leaves dull green in colour, flowering and fruiting are sparse.

Application of phosphorus 44.8 kg/ha increased the seed yield of *C. capsularis* and *C. olitorius* (Das and Kundu, 1966). It is reported that in jute seed production neither source of phosphatic fertilizer nor method of application produced any significant difference..

Higher increased in seed yield were obtained from the application of nitrogen + phosphorus than either of them alone (Hussain *et al.,* 1975). Dasgupta (1989) opined that two jute cv. JRC-321 and JRO-534 with increase the level of phosphorus i.e. 30-40 kg P_2O_5/ha gave higher seed yield of jute and its growth. Sinhamahapatra (1991) reported that basal application of phosphorus 40 kg/ha gave the higher seed yield.

C. olitorius has been reported to show a greater response to K than *C. capsularis* (Gupta and Sen, 1953). With K deficiency, the internode length is reduced, the leaf margins develop brown, necrotic spots, the plants become susceptible to disease and seed quality is poor (Gupta and Sen, 1953; Sen and Banerjee, 1960). Crop requires a basal dressing of potassium @ 15 kg/ha to produce healthy seed (JARI, 1973). Seed yields of jute have been significantly increased by application of potassium @ 44.8 kg/ha (Das and Kundu, 1966).

Chowdhury (1986) from a pot experiment conducted over two years with *C. capsularis* (var. JRC-212) and *C. olitorius* jute (JRO-632) reported that 40 kg K_2O/ha influenced growth and yield of both jute cultivars. Still higher levels of 80 kg K_2O/ha exerted adverse effect.

Das (1995) reported that application of potassium 40 kg/ha increased seed yield (734.9 kg/ha) of jute (*C. olitorius*) by improving the yield attributing character is no. of pods/plant, no. of seeds/pod and thousand seed weight. Application of potassium increased seed yield of jute irrespective of sowing date or clipping operation over control (Annual Report, 1973-74).

Dalai *et al.* (2001) observed that application of 40 kg K_2O/ha, but significant difference was obtained only in thousand seed weight. Maximum and significantly seed yield of jute (741.2 kg/ha) was obtained at 40 kg K_2O/ha.

Potassium (40 kg K_2O/ha) with 40 kg N, 30 kg P_2O_5 and 20 kg S/ha significantly increased the seed yield of both *C. capsularis* and *C. olitorius*, but the magnitude of increase of seed yield was more in case of *C. capsularis* than *C. olitorius*. The maximum seed yield was obtained with higher interaction effect of potassium (40 kg K_2O/ha) and sulphur (40 kg S/ha) application (Mondal *et al.,* 1998-99).

It has been observed that a response of 0.1 q/ha of seed yield of jute was obtained from the average of six location in Bangladesh with application of 20-30 kg S/ha. Das (1995) reported that all yield attributing characters i.e. number of branches per plant, number of pods/plant, number of seeds/pod and thousand seed weight improved significantly due to application of sulphur 20-40 kg/ha or resulted significantly higher seed yield of jute over zero level of sulphur.

Mondal and Dalai (1998) observed that with increasing level of sulphur upto 40 kg S/ha number of branches per plant and number of seeds per pod increased significantly over lower levels of sulphur. Thousand seed weight also increased but difference was not significant over 20 kg S/ha. The maximum yield was obtained at 40 kg S/ha along with recommended dose of N, P and K; it was significantly superior to lower levels of sulphur.

5.1.6.6 Water management

Jute for seed production is mainly a crop of monsoon season. So, it suffers unpredictable heavy rainfall resulting water-logging or exposes to drought at early stage of growth. Adequate soil moisture should be maintained before the onset of monsoon. Irrigation should be given as and when required. Generally, two to three irrigations may be needed in early sowing crop at an interval of twenty days during the growth period depending on rainfall. various authors studied the effect of irrigation interval on seed yield of two varieties of jute after flowering. Irrigation was carried out at 20 days after sowing as required and subsequently at 7, 12 and 21 days whereas after flowering. Seed yield increased progressively with increasing frequently of irrigation. Yields will always higher in *C. olitorius* than *C. capsularis*.

5.1.6.7 Decapitation

Pruning of apical portion of plant is called decapitation (Syn. Detopping or clipping). Pruning or decapitation induces branching; under favourable condition such branches grow vigorously and lead to an increase of about 10 per cent seed yield. Pruning or clipping is done once, when the seed crop grows vigorously and attains height of 45 to 50 cm in 45 to 48 DAS. Care should be taken to ensure that soil is not deficient in nutrient and moisture during the operation. The soil must be fertile or sufficiently fertilized to support a good growth of the branches following loss of the apical bud; clipping should be avoided in late sown crop (Nath, 1980).

JARI in its annual report, 1986 clearly stated that the treatment constituting line sowing and detopping gave 15% higher seed yield than line sowing without detopping. There should be enough nutrients in the soil and sufficient space to promote full growth of at least four branches which are to reach flowering stage; such plant yield more seeds (JARI, 1973).

Under certain situation pruning of the jute plant induces good branching and leads to a better seed yield. But there is not a universal recommendation. It should be mentioned that pruning is not to be done under conditions described below.

(i) An area where rainfall during the vegetative period is less than 800 mm and number of rainy days is below 30.

(ii) Where crop growth is poor, such as one finds in 'Sandy',

(iii) When nutrient status of the soil would not permit renewal of growth after pruning.

In such cases pruning may lead to loss of seed (JARI, 1977). It has been reported that after July, clipping is not advisable. Clipping upto 3rd week of July yielded more seed than non clipped ones.

It has also been observed that decapitation at 45 DAS had also positive effect on seed yield of jute. Yield attributing characters like average number of branches and fruits per plant were improved by increasing spacing and decapitation.

Dasgupta (1989) reported that significantly highest seed yield was obtained at 3rd June sowing with pruning in both the species of jute. The yield attributing characters like number of branches per plant, number of capsules/pod and weight of 1000 seeds significantly increased and influenced on seed yield. Das (1995) reported that decapitation of jute (JRO 524) at 45 days after sowing significantly improved all yield attributing characters except thousand seed weight and gave significantly superior seed yield (11% more) than without decapitation. Dey *et al.* (1998) reported that pruning of plants on increasing the seed yield of jute and higher net returns an experiment was conducted West Bengal 1991 at Nadia, Pruning of apical meristems increase the number of branches per plant ultimately increased the seed yield. (Mondal *et al.* 2003).

5.1.7 Seed Production by Cuttings

Jute cuttings from the apical portion strike root easily under humid condition. Cultivators who want to retain their own seed by this method just before the harvest of jute for fibre 1-9" apical portion is cut off and transplanted in the well prepared land during early August. The seeds can be obtained from these cuttings without loosing the fibre yield (Mondal *et al.* 2007). Roy (1964) studied the jute seed multiplication by vegetative propagation (cuttings of jute stem) and observed that the fairly high yield per cutting (5.27 g in JRC-212 and 4.83 g in JRO 632) which is reflected in the ultimate total yields of these varieities and the high germination percentage of the seeds in either case indicate the possibility of getting a high yield of seed of normal quality of cuttings of jute stems which are about three months old and have not yet flowered and planted towards the middle of August. The estimated rate of seed production from JRC-212 cuttings (6.5 q/ha) compared favourably with the average seed yield (6.8 q/ha) of this variety when raised directly for seed. It was also reported that the seed yield of capsularis jute is about 4 to 5 q/ha. Further, the jute seed yield from the cuttings of JRO 632 is higher than that previously reported for this variety by following the usual method of seed multiplication. Thus, from a trial conducted for three years the average seed yield of this variety has been shown as 4.4 q/ha, while the seed yield for olitorius jute in general by the usual method was reported to be 3 q/ha. The number of seeds per gram is 303 in JRC-212 and 538 in JRO-632 which compare favourably with 280 per gram for *C. capsularis* and about 500 for *C. olitorius* produced by the usual method. This shows that the seeds of cutting were as good and developed as those produced by the usual method.

Roy (1963) studied the jute seed multiplication by vegetative propagation when the jute plants were about 3 months old and had not yet came into flowering, cutting of stems (about 30 cm) of *C. capsularis* (JRC 212) and *C. olitorius* (JRO 632) were planted, 15 cm apart, in well moist soil in rows with 30 cm spacing in between. The lower end of the cuttings was inserted about 7.5 cm to 10 cm deep into the soil in an inclined position. Proper management practices were carried out at regular intervals. Four weekly after the planting the cuttings were top-dressed with ammonium sulphate (170 kg/ha). Those of the cuttings that set well developed a large number of roots and the fruits looked well formed as with case of plants rainsed directly from seeds. The vegetative method of seed multiplication has the following advantages (Roy, 1963; Mondal *et al.,* 2002; Mondal *et al.,* 2007).

(i) For instance, when the fibre crop of both capsularis and olitorius varieties is harvested early so as to release the land for transplanted *Kharif* rice with double cropped areas, the top section of the harvested plants may be planted either in the form of cuttings or toppings and profitable utilized for seed multiplication. Jute fibre and seed will thus be obtained from the same plant and the cost of seed multiplication will be reduced.

(ii) The method can also be very useful when quick multiplication of a limited stock of mother seed is needed. Since, several cuttings may be obtained from single plant and the seed yield of several cuttings will out weigh that of a single plant.

5.1.8 Plant Protection Measures

Plant protection measures should be taken in advance to save the seed crop from insect pests and diseases. Agarwal (1980) also reported that for producing disease free jute seeds, it is essential that necessary precautions are taken right from the very beginning to control the incidence of disease and pests.

5.1.8.1 Insect

Jute is attached by large number of insect pests, mainly semilooper, hairy caterpillar, stem weevil, mite (red and yellow) etc. Three to four spraying with endosulfan (0.04 to 0.07%) during the cropping season can be applied to control major pests. These may be applied to control major pests. These may be applied at 20-25 days intervals starting 35 to 40 days crop age. To control yellow and red mite dusting lime sulphur (3:1) 20 kg/ha is effective. Spraying with Kelthane 0.04% is also very sound in efficient control of red mite in *C. capsularis* whereas endosulfan 0.04% is effective against yellow mite.

5.1.8.2 Diseases

The following general control measures may be taken against various diseases of jute.

(i) Jute seed should be treated with fingicides against seed-borne pathogen @ 3 g/kg of jute seed. The seed should be thoroughly mixed in a rotary drum for 10 minutes.

(ii) In soil which are acidic, liming @ 3 to 4 tonnes per hectare may be done in every 3 to 4 years depending on conditions.

(iii) Application of potassium @ 35-40 kg/ha depending on requirement helps to reduce the disease infection.

(iv) Timely weeding and thinning to maintain proper spacing and clean cultivation ensures disease free crop.

(v) Improvement of drainage condition.

(vi) Proper crop rotation should be followed; potato-jute rotation induces the *Macrophomina phaseoli*. Introduction of cereals instead of potato is helpful to check the disease.

(vii) Uprooting and burning the infested plants.

Table 5.3: Specific control measures for particular diseases are given

Disease	Control measures
Stem rot and Root rot (*Macrophomina phaseoli*)	Seed may be treated with Bavistin or Dithane-M-45. Two to three times spraying with Dithane-M-45 (0.2 to 0.3%) during the crop season to control the disease.
Anthracnose (*Colletotrichum gloesporioides* on Tossa jute and *Colletotrichum corchorus* on white jute)	Spray the jute crop with Copper oxychloride (0.75) of 7 days interval. Lime sulphur (3:1) dusting completely control the disease. Dusting should be done in the morning hours. Prophylactive spraying of Bordeaux mixture can control the disease.

5.1.9 Harvesting and Threshing

Harvesting of jute seed crop is done at full maturity but before the pods start shattering. The jute seed crop is ready to harvest in October – November when seed pods become hard and brown in colour. The good quality of seed of *C. capsularis* can be obtained when the crop is harvested before dead ripe of stems and pods. Several hand cuttings may help to fulfill this method. But in case of *C. olitorius* the harvesting should be done at dead ripe stage of stem and pod. The matured seed crop should be harvested with the help of sickle in morning time by minimum bending and jerking the plant. The jute plant bundles are kept for sun drying on cementing or concrete threshing floor and threshing may be done by beating pods with the help of a wooden stick. Necessary precautions should be taken to avoid contact with moist soil, mechanical admixture and broken the seeds. The seeds, then winnowed and should be sun dried until the moisture content of seed is below nine percent. Then this healthy seeds are kept for storage.

5.1.10 Seed Yield

On an average 3-4 q/ha of jute seed is obtained in case of *C. olitorius* and 4-5 q/ha in *C. capsularis* jute. Das (1995) reported that higher seed yield (759 kg/ha) of *C. olitorius* was obtained with the application of 20 kg N, 30 kg P_2O_5, 40 kg K_2O and 20 kg S/ha with decapitation at 45 days after sowing of jute. It has been reported that higher seed yield (in *C. capsularis* 1032 kg/ha and in *C. olitorius* 717 kg/ha) was obtained in May – sown crop and the crop was fertilized with 15 kg N, 15 kg P_2O_5 and 15 kg K_2O, maintained the spacing 30 cm with decapitation. Thakeria, and Sharma (1991) reported that the higher seed yield (1.16 – 1.18 t/ha) was obtained from *C. capsularis* jute when sown in 1st June to 16th June with a spacing of 30 cm x 10-15 cm. In June sowing variety JRC-7447 yielded highest (6.56 q/ha) followed by JRC-212 and JRC-321 (Annual Report 1982, JARI). Maximum yield of jute (769 kg and 875 kg/ha in olitorius and capsularis respectively) was obtained when jute crop fertilized with 40-60 kg N, 40 kg P_2O_5 and 40 kg K_2O and 40 kg S/ha. Production Potential of Jute seed was influenced by Potassium, Sulphur and decapitation (Dalai *et al*, 2001; Mondal *et al.* 2002)

5.1.11 Storage of Jute Seed

The jute seed immediately after harvest contains on an average 21.4 per cent, after 3 days of sundrying it falls further to 8.8 per cent and 7.3 per cent average after four days. Hence, proper sun drying, at least 4 days, followed by treatment with organo mercuric compounds (3 g/kg of seed) is recommended before storage. Storage is best under dry; air tight conditions. The seeds are required not only to maintain their viability in storage but also to preserve their physiological stamina for germination under different field conditions and production of good crop. Storability of a bulk seed depends on a combination of factor i.e. moisture content, temperature, degree of invasion of storage fungi, length of time seed is to be stored and the last but not the least the genotype of the seeds. For preservation of the seeds in bulk, storage in double plastic bags with silica gel at ambient temperature and relative humidity is advantageous because silica gel can be recovered by healing at 125°C for 2 hours and it can be used suitably in the case of unsold seeds which have gained moisture due to storage in cotton bag.

5.1.12 Viability and Vigour Test of Jute Seed

The viability of jute seed can be tested with fair accuracy with tetrazolium compound and by germination test conducted on moistened blotting paper. This quick and simple methods have been developed at IJIRA for examining the viability of jute seeds.

(i) *Seed crushing test :* Moistened seeds are directly crushed on the blotting paper and the spots left behind are examined under natural light. Viable seeds show yellow spots while the non-viable seeds show yellowish orange spots.

(ii) *Aqueous test :* Seeds are soaked in excess of water for 3-4 hours and the supernatant liquid is used for observation. The aqueous extract of viable seeds is colourless while that of non-viable seeds is yellow to yellowish orange.

Khandakar (1980) suggested that germination tests should be carried out for 4 days and then terminated. The vigour was calculated with the formula

$$= \frac{\left(\frac{a}{1}+\frac{b}{2}+\frac{c}{3}+\frac{d}{4}\right)}{S} \times 100$$

Where a, b, c and d are the number of seeds germinated at 30°C in duration after 1, 2, 3 and 4 days respectively and 'S' is the total number of seeds germinated after four days. Root and shoot length were measured after 24 hr for 4 days. For seed germination test 95% confidence limits were calculated following Roberts (1963) containing table.

References

Agarwal, R. L. 1980. Seed Technology. Oxford and IBH Publishing Co., New Delhi.

Anon, 1956. Jute – world's foremost bast fibre. Economic Botany, 10 : 103-33.

Banerjee, K. and Rakshit, S. C. 1970. Use of male sterility in the production of heterotic hybrid jute (*C. capsularis* L.). Indian J. of Agric. Sci., 40(8) : 691-696.

Basak, S. L. and Chaudhuri, B. B. 1966. Extent and nature of natural cross pollination in Tossa Jute (*Corchorus olitorius* L.). Indian Journal of Agricultural Science, 36 : 267-272.

Basak, S. L. and Gupta, S. 1972. Quantitative studies on the maturing system of jute (*Corchorus olitorius* L.). Theoretical and Applied Genetics, 42 : 319-324.

Basu, K. O. 1977. Acta Hort. Culture No. 53 : 355-73.

Bhaduri, P. N. and Bairagi, P. 1969. Relation between vegetative and sexual compatibilities among jute species. Sci. Cult. 105-108.

Bhaduri, P. N. and Chakraborty, A. K. 1948. Colchicine induced autotetraploid jute *Corchorus capsularis* and *C. olitorius* and the problem of raising improved varieties. Sci. Cult., 14 : 212-213.

Chaudhuri, N. C. 1933. Jute and substitutes. Calcutta, Newman and Co., 3rd edn., pp. 249.

Chaudhary, J. 1986. Studies on the effect of potassium on growth and yield of olitorius jute. Indian Agriculturist, 30(4): 275-280.

Das, T.K. 1995. Seed Production of jute (*Corchorus olilorius* L.) influenced by potassium, sulphur and decapitation M.Sc Thesis.

Dasgupta, U.K. 1989. Studies on production potential of jute seed as influenced by non-monetary inputs. Date of sowing and Dicapitation, M.Sc. Thesis.

Chowdhuri, S. D., Eunus, A. M. and Jabbar, M. A. 1956. Studies in Colchicine induced tetraploid of jute. Euphytica 5 : 339-346.

Dalai, G.M. Mondal, S.S. and Maity G.C. 2001. Production potential of jute and as influenced potassium, sulphure and decapitation for sustainable agriculture. Natioanal Seminar oon Frontiers of crop Management. pp 84-86.

Das, B. K., Dua, D. B., Mitra, S. N. and Sarkar, A. 1959. Proc. Indian Science Congress Paer-III, 471.

Das, B.K. Kundu, B.C. 1966. Response of varieties to complete fertilizer on the yield of jute seed. Indian Agric. 1966. 10,52-58.

Datta, R. M. Investigation on the autotetraploids of cultivated and wild types of jute (*C. olitorius* L. and *C. capsularis* L.) Ziichter, 33 : 17-33.

Dey P., Banerjee, N.C. and Bhattacharya, S.P. Das, D. and Bhattacharya, M. 1998. Environment and Ecology 16(1): 1-6

Doraiswamy, I., Basu, A. and Chellamani, K. P. 1998. Proc. Int. Seminar on Jute and allied fibre – changing Global Scenario, 5-6 February, 1998, pp. 7-10.

Dutt, N. and Ghose, S. K. 1962. Measurement of natural crossing as effected by insects in olitorius and capsularis jute. Indian Journal of Agricultural Science, 32(3) : 242-250.

Gupta, Sen, P. 1959. Manuring of Jute in relation of soil category, J. Indian Society Soil Science 1959. 7, 97-101

Finlow, R.S. and Burkill, I.H. 1912. Inhertance of red clour and the regularity of self fertilization in *C. capsularis* linn., The common jute plant. Merm. Dept. Agric. India. Bot. 4: 73-92

Finlow, A., Howard, G. L. C. and Khan, A. R. 1919. Studies on the pollination of Indian crops. Mem. Dep. Agric. Indian, Bot., 10 : 195-220.

Ganesan, A. T., Shah, S. S. and Swaminathan, M. S. 1957. Cause for the failure of seed setting in the cross *Corchorus olitorius* and *Corchorus capsularis*, Current Science, 26 : 292-303.

Ghose, R. L. M. and Dasgupta, B. 1945. Floral Biology, anthesis and natural crossing in jute. Indian Journal of Genetics, 4 : 80-84.

Ghosh, N. and Sen, S. 1974. Isolating distance in Tossa jute. The Society for the Advancement of Breeding Research in Asia and Oceania. Sabrao Journal, 91-92.

Hossain, G., Amin, A.M. Sabemi, R. and Ahmed, B. 1975. Effect of nitrogen and phosphorus on seed yield of jute J. of Agricultural Research. 13(4): 711-716.

Howrad, A., Howard, G.I.C. and Khan, A.R. 1919. Studies on the Pollination of Indian Crops. Mem. Dept. Agric. India Bot. 10: 195-220.

JARI, 1958-59. Annual Report, Jute Agricultural Research Institute, Barrackpore

JARI, 1959-60. Annual Report, Jute Agricultural Research Institute, Barrackpore.

JARI, 1973. Annual Report, Jute Agricultural Research Institute, Barrackpore.

JARI, 1977. Annual Report. Jute Agriculture Research Institute, Barrackpore, West Bedngal.

Joseph, J., Kundu, T.K. and Sarkar G.R. (1970). Influence of sowing time on seed yield of while jute (*Corchorus capsularis* L). Indian J. Agric Sci. 40(10): 883-888.

Jute Agriculture Research Institute, 1973. Annual Report prepared by V. N. Saraswat and M. K. Pal.

Kar, B. K. 1959. Proc. 46th Indian Science Congress, pt 2, Presidential Address, 166-184.

Kundu, B.C. 1956. Jute-word's foremost fast-fibre. I Economic Bot. 1956. 10, 103-33.

Kundu, B. C. 1951. Origin of Jute. Indian J. Genetics and Plant Breeding, 11 : 95-9.

Kundu, B. C. and Rao, N. S. 1956. Studies on cochicine induced tetraploids of *Corchorus olitorius* Linn. (Jute). Journal of Indian Botanical Society, 35 : 11-26.

Kundu, B. C. and Sarma, M. S. 1956. Journal of Indian Botanical Society, 15 : 11-26.

Kundu, B. C., Basak, K. C. and Sarkar, P. B. 1959. Jute in India (Monograph) Indian Central Jute Committee, Calcutta.

Mian, A. L. and Gain, M. D. 1971. Indian J. Agricultural Science. 41(11) : 938-43.

Mondal S.S., Dalai, G.M. and Ghosh, A. (2002) Studies on influence of K. S. and non-monetary inputs on nutrient uptake fertility build up of soil in soil under jute seed Production in alluvial zone of W.B. Abstract of 6th Agricultural Sceience Congress, Bhopal, India pp. 45.

Mondal *et al*. 2007. Effect of potassium and sulphur on seed production of olitorius jute raised from top cutting. Environment and Ecology, 25(3) : 484-486.

Mondal, S.S., Brahmachari, K. and Maiti, S. 1999. Studies on effect of K. S and organic manure matter on the yield and quality of Jute fibre under intensive cropping. Proceeding of the International Seminar Jute and Allied Fibres-changing Global Scenario on Jute and Allied Fibres – Changing Global Scenario 5-6 February, 1998. pp 99-103.

Mukherjee, D. K. 1948. Effect of sowing time on growth and development of jute. Pt. 2. Indian Journal of Agricultural Science, 18 : 129-136.

Nakajima, G. 1949. Tetraploid plants of jute and china jute raised by colchicines method. Jap. J. Gent. (Suppl.) 2 : 121-131.

Nath, S. 1980. Grow jute for seed. Seeds and Farms VI (2).

Patel, J. S. and Datta, R. M. 1960. Interspecific hybridization *Corchorus olitorius* Linn. And C. capsularis Linn. And the cytogenetical basis of imcompatibility between them. Euphytica, 9 : 89-110.

Ray, Bhaswati and Majumder T.K. (1995). Effect of Sowing rate on niterogen level on Seed yield of white jute and Tossa Jute. Indian J. of Agri. Sci. 65(12): 891-893.

Rao, K. R., Sanyal, A. T. and Datta, J. 1944. Colchicine treatment of jute. Sci. Cult., 10 : 86-89.

Roy, B. 1964. Correlation between plant height and flowering time of jute (*Corchorus olitorius* L.). Jute Bull. 27,No. 2, 32-3

Roy, B. 1960. Natural crossing in *Corchorus olitorius* L. Indian Agriculturist IV(1) : 1-4.

Roy, R. K. and Singh, K. N. 1988. Jute Development J. 8(4) : 18-20.

Saraswat, V.N. and Pal. M.K. 1984. Package of Practice for Jute Seed Production. Field Bulletin on J.A.. 1.2 Jute Agricultural Research Institute (ICAR) West Bengal, 5.

Saraswat, V. N. and Pal, M. K. 1972. Package of practices for jute seed production. Pub : Jute Agricultural Research Institute, Barrackpore, pp. 1-4.

Sen, P.K. and Banerjee, K.L. 1960. Physiological Studies in Jute. Effect of nitrogen phosphorus and potassium on the growth of jute (*Corchorus olitorius* L. var E-G). Indian Agric. 1960. 4:121-8.

Sinhamahapatra, M.K. 1991. Seed Production of Jute influenced by sowing dates, decapitation and foliar application of mepiquat chloride M.Sc. Thesis. Bidhan Chandra Krishi Viswavidyalaya, Mohanpur Nadia, West Bengal.

Singh, A. P., Ganguli, O. M. and Sasak, S. L. 1984. Seed Res. 12(2) : 87-91.

Srinath, K. V. and Kundu, B. C. 1952. Cytological studies on pollen tube growth in receiprocal crosses between *Corchorus capsularis* Linn. And *Corchorus olitorius* Linn. Cytologia, 17 : 219-223.

Thakuria, K. and Sharma, K.K. (1991) Effect of sowing dates and spacing on seed yield of Capsularis Jute. Indian J. of Agronomy. 36(1): 36-39.

5.2 Cotton

Cotton is an important commercial crop to the farmers. Cotton as fibre crop is cultivated in many countries of the world. The USSR, USA, China, India, Brazil, Pakistan, Turkey, Egypt Mexico and Sudan are the main country can produce 85% of the total production. The textile industry is very crucial to the Indian economy. The sector is the second largest employment provider after agriculture, employing over 38 million people directly and another 50 million indirectly. It is the third largest producer of cotton and cellulose fibre/ yarn, second largest producer of cotton yarn. The area under cotton is India is more or less constant ranging from 8.5 to 9.0 million hectares. On the other hand, productivity during the same period increased by four fold from 104 kg lint/ha in 1956-57 to 462 kg lint/ha in 2004-05. Production increased to 24.3 million bales in 2004-05. By way of exports, foreign exchange earnings of cotton amount to about Rs. 50,000 crore, which is nearly one-third of the total foreign exchange earning of our country.

5.2.1 Origin and Distribution

The true cotton, the list bearing species, belong to the genus *Gossypium* of family Malvaceae. The genus *Gossypium* according to Fryxell contains 39 species of which 4 are cultivated. Based up on the chromosome number, the species earn also be distinguished as diploids (n=13) or tetraploids (n=26). The diploid species are also called as Old World or Asiatic cottons and comprise the species *G. arboreum* and *G. herbaceum*, *G. barbadense* and *G. hirsutum* have 26 pairs of chromosome and distinguished as the New World cottons. The new world cottons are natural amphidiploids containing A genome from a taxon of the Asiatic diploid group and a D genome from a taxon of the American diploid group. The new world cottons are popularly known as American (*G. hirsutum*) and Egyptian (*G. barbadense*) cottons.

Gossypium hirsutum is the principal cultivated cotton and accounts for about 90% of the world cotton production. It originated in southern Mexico and Central America as a perennial shrub but is cultivated as an annual. In the United States it became known as American Upland cotton because it was cultivated on higher land, inward from the low coastal areas of Georgia and the Carolinas. American Upland cotton produces fibres that vary from short to long according to the variety being grown.

Gossypium barbadense originated in Central and South America and has long, fine fibres. A type known as Sea Island was formerly cultivated along the Coastal Islands of South Carolina and Georgia, but none is grown there at the present time. It was introduced to the Nile valley of Egypt, where it became known as Egyptian cotton and was prized for its long, strong staple. Currently grown in irrigated areas of Arizona. In New Mexico and western Texas, it is generally referred to as American Pima or more commonly Pima. The varieties of Pima cotton grown in the United States are normally later in maturity than the American Upland varieties. *Gossypium barbadense* accounts for about 8% of the world cotton production.

India is only country in the world where all the four cultivated species of cotton are grown on a commercial scale besides hybrids (both inter and intra specific). The maximum area is covered by hybrids (40%), followed by *G.*

hirsutum (36%), *G. arboreum* (16%), *G. herbaceum* (8%) and *G. barbadense* (0.2%). In the United States 99% of the area is sown with *G. hirsutum* and remaining is under *G. barbadense*.

5.2.2 Botany

The cotton has primary tap root system growing straight down to a depth of 1.8 m or more and lateral root confined to the top 60 cm. The root development continues until the onset of flowering, about 8-10 weeks after planting. It has the main stem which follows the compound interest low falls off soon after the initiation of the first flower. At the axil of leaf, one branch is there, which is of two types (i) Vegetative or monopodial and (ii) Reproductive or sympodial

Fruiting branches result from the transformation of vegetative meristem into flower primordium. The floral differentiation consumes the meristem of the axis and thus the axis terminates with the flower, with the extinction of terminal growth by the floral development, the first axillary primordium at the base of the true leaf is stimulated.

Cotton has a complex branching habit because : (i) both the axillary and extra axillary buds may be vegetative or reproductive (ii) any axillary may be dormant or active and (iii) each segment of a sympodium is a new first axillary axis.

The petiolate and stipulate leaves are spirally arranged on the main stem and vegetative branches. The petioles and leaves are either glabrous or pubescent or fully hirsute depending on the variety. The leaves are generally cordate with palmate lobing.

5.2.2.1 Floral biology

The cotton flowers are surrounded by three triangular bracts forming what is commonly known as squares. On the day preceding pollination, the twisted corolla emerges from the square. In American Upland varieties, the corolla is cream colored when the flower first opens, but turns red the second day and later falls from the plant. The pistil is formed normally from three to four carpels in Pima cotton and four to five carpels in American Upland cotton. The number of locks in the boll corresponds to the number of carpels comprising the compound pistil. Normally, seven to nine seeds are set within each lock. The stamens form a tube that encloses the styles and is united with the inside base of the corolla. Pollen is shed directly on the stigma when the anthers open, or it may be carried to the stigma by insects.

5.2.2.2 Pollination

Pollen is wind-borne only to a very slight extent, if at all, an account of its heavy sticky nature. Cross-pollination from pollen being carried by insects, mostly bees and usually varies from 5 to 30%, although amounts in excess of 50% have also been reported. It leads to increase in heterozygosity in the cotton plant. This results from natural crosses between genotypes within the variety. Varieties of cotton are rarely if ever pure lines as in small grains or soybeans. A moderate degree of heterozygosity is generally desired in a cotton variety to provide some heterosis and to maintain maximum yield and adaptation.

Table 5.4: Morphological differences in the cultivated species

Plant part	*G. arboretum*	*G. herbaceum*	*G. hirsuum*	*G. barbadense*
Vegetative branches	If present, ascending, thin and flexible	Few	None to many and horizontal	Few to many, horizontal
Fruiting branches	2-jointed	Many jointed	Many jointed	Many jointed
Leaves	Two-third to four-fifth cut into 5-7 lobes	Half cut or less into 3-7 lobes	Cordate half or less cut into 3-5 lobes	Two third cut into 3-lobes
Lobes	Ovate, oblong or curvilinear	Ovate, rotund to rounded	Broadly triangular acuminate, lateral lobes divergent	Long, tapering acuminate, slightly constricted at base
Stipules	Linear or falcate	Small, linear	Falcate or auriculate, caduceus or persisting	Variable in size, falcate or auriculate
Bracteoles	Closely investing the bud, more or less triangular, usually longer than broad, 3-4 teeth at apex	Flaring widely from flower and capsule broader than long margin serrated into 6-8 broadly triangular teeth.	Longer than broad with 4-12 long acuminate teeth	As long as broad, cordate, gashed into 10- 15 long acuminate teeth
Petals	Pale corolla common	-	Narrow tube, widely expanding	Large, corolla not widely expanding.
Staminal column	Long antheriferous	Anteriferous throughout anther, filaments short	Short, anthers loosely arranged upper anther filaments longer and ascending than lower ones	Long anthers closely packed
Stigma	United, rarely divided at top	United throughout	United throughout	United at the top, if clefted never spreading
Capsules	Tapering, profusely pitted with oil glands usually 3 rarely 4 locular with 6-17 seeds	Rounded, beaked, 2-3.5 cm long, few oil glands 3-4 locular, not more than 11 seeds locule	Variable, rounded or tapering, few or inconspicuous oil glands 3-5 locular, 5-11 seeds locule	Variable, more than 3-7 cm long, broadest near the base coarsely fitted 3-z locular with 5-8 seeds locule

5.2.2.3 Male sterility

Several male-sterile genes have been identified in cotton which may be utilized in the production of hybrid cotton. At least one recessive male-sterile gene, ms2, normally gives complete sterility. Of the others male-sterile genes, several give only partial sterility and some are dominant. The recessive male-sterile gene would have been utilized in facilitating cross-pollination, although male sterility may be unnecessary due to the natural cross pollination of cotton flower.

Cytoplasmic male sterility has been produced by transfer of *G. hirsutum* or *G. barbadense* chromosomes into *G. harknessii* cytoplasm. Fertility restoration is obtained from a single partially dominant gene (Rf) transferred from a restoration is obtained from a single partially dominant gene (Rf) transferred from a *G. harknessii* chromosome. The gene gives good fertility restoration in commercial upland varieties when homozygous, but only fair fertility restoration when heterozygous. This causes problems in cotton. Fertility restoration is improved by presence of a dominant fertility enhancer gene (E) from pima cotton. Cytoplasms from *G. anomalum* and *G. arboreum* with *G. hirsutum* chromosomes are cytoplasmic male sterile, but the make sterility is not environmentally stable.

5.2.3 Hybrid Cotton

The hybrid vigor in cotton is primarily utilized for higher production of lint fibre, early maturity, adaption to mechanical harvesting, resistance to stress environments, resistance to disease and insect injury and improvement in fiber and seed quality. Tools used for producing F1 hybrid seed are hand pollination, natural and use of genetic on cytoplasmic male sterility. The production of commercial hybrid seed by hand emasculation and pollination is commonly practiced but it is a labour intensive methods.

The A- and B-lines in a particular hybrid have identical genotypes, but the A-line is in *G. harknessii* cytoplasm and is male sterile and the B-line is in normal. *G. hirsutum* cytoplasm and is male fertile. The A- line is maintained by pollination from the B-line. Hybrid seed is produced by pollination of the A-line by an R-line that contains fertility-restoring genes. There is some evidence that fertility restoration varies slightly with the genotype of the A-line, suggesting that A-lines may differ in modifying genes that influence fertility restoration. In addition to restoring fertility, the R-line needs to make production of hybrid cotton attractive to the grower, or improve other characteristics of the hybrid.

Some important points are to be kept in mind for successful hybridization programme in cotton.

(i) To make the system economically viable high percentage of seed set is necessary.

(ii) Gross-pollination ensures the need.

(iii) Populations of natural insect pollination such as wild bees are generally insufficient to cross pollination.

(iv) It seems that cotton flowers are not attractive to honey bees, honey bees are not good foragers for cotton pollen

(v) Flowering of the R line will overlap that of the A line.

5.2.4 Climate

Cotton, a semi-xerophyte, is grown in tropical and sub-tropical conditions for fibre production. The seed crop is usually sensitive to the vagaries of weather. Continuous rains or a long spell of dry weather may prevent the sowing of the crop at the proper time, hinder seed germination, or retard crop growth. Heavy rains, continuous high winds or excessive drought at flowering and fruiting time may cause heavy shedding of buds and young bolls, thereby ruin the crop. An early frost may kill the plants permanently. Ultimately rains and a heavy humid weather during the later stages of the cotton season may spoil the produce, lower its spinning properties in fibre production or promote the attack of pests and diseases and detrimental effect on seed quality in seed production.

Climatic requirements for the successful seed production of cotton

i) A mean annual temperature of over 60°F

ii) An annual rainfall of at least 200mm with favourable distribution

iii) Abundant sunshine during the period of boll maturation and harvesting of the produce

iv) A frostless season of 180-240 days in North India. The cotton plant can not stand frost and its cultivation is restricted largely to flat or undulating condition upto an altitude of 3000 ft.

For good plant growth, cotton generally requires a rather high and uniform temperature. The optimum temperature range for vegetative growth is 70-80° F. During the period of fruiting day temperature of 80-90° F and cool nights are needed for the best results. For harvesting of good quality seed, bright sunny days are essential.

5.2.5 Selection of Land

Medium to high low situation with good drainage is suitable for cotton seed production, because it could not tolerate water-logging condition. Land for cotton seed production must be free volunteer plants. Cotton can be grown on variety of soils, but the soil should be deep, fertile and moisture retentive. Cotton is grown in black cotton and medium to heavy black soil as well as in the alluvial soils as irrigated crop. Red laterite soil is also suitable for cotton cultivation. Research findings also revealed that the soil should be free from hard layer of carbonates and hard pan in the sub-surface layer which hampers the growth and developments of plants and ultimately negatively influences seed production.

5.2.6 Isolation Distance

To maintain the genetic purity for seed production of cotton isolation distance depends on the extent of natural cross pollination. Cotton is a self-pollinated crop, but natural cross pollination occurs to some extent in *G. hirsutum* (10 to 50%), *G. arboreum* (1 to 2%) and *G. barbadense* (5 to 10%). For seed production, it should be better to grow only one variety of cotton for each seed multiplication form. The isolation distance for foundation and certified seed production are fifty and thirty inches, respectively, should be maintained from the fields of other varieties of same or other species and fields of same variety not conforming to the varietal purity. FAO (1961) indicated the isolation distance (Table-5.5), where diseased rows used for cotton seed production of different classes.

Table 5.5: Number of diseased rows for different classes of cotton seed production

Classes of seed	Size of field	Distance between varieties	Number of diseased rows
Breeders and Foundation Seed	Any size	Less than 45 m	20
		45 to 90 m	10
		Over 90 m	0
Registered	Less than 8 ha	Less than 45 m	15
		45 m or more	0
	8 ha or more	Less than 27 m	5
		27 m or more	0
Certified	Less than 8 ha	Less than 27 m	10
		27 m or more	0
	More than 8 ha	0	0

Source: Agricultural and Horticultural Seeds, FAO, Agricultural Studies (1961), No. 55

5.2.7 Cultural Practices

5.2.7.1 Land preparation

In summer, one deep ploughing should be given with soil turning plough followed by three to four harrowings. After each harrowing laddering is essential to make soil pulverized and leveled. Harmful weeds, offtype plants and stubbles should be eradicated from the field before sowing the crop.

5.2.7.2 Planting variables

Sowing time

The best time of sowing of cotton for seed production in April-May as irrigated crop. The rainfed seed crop is sown in June – July, with the commencement of the monsoon. In southern part the cotton is sown September – October. It was found that pre-monsoon sowing was better than monsoon sowing. It was also reported that crowbar planting and polythene tag transplanting were observed economically beneficial. It was also established that 1st May planting showed maximum plant height, bolls/plant, boll weight and markedly reducing boll shedding as well as seed cotton yield was the highest at this sowing date. Delay in sowing by every 15 days not only significantly increase the boll-shedding intensity but also decreased the seed cotton yield. On an average sowing of cotton on 16 and 31 May reduced the yield by 13.1 and 31.9% respectively compared with that sown in 1 May.

Recently released varieties / hybrids of cotton

Variety	Duration (days)	Yield (t/ha)	Area of adoption	Salient features
GN Cot. Hy-18	150–160	2.36	Irrigated conditions across south Gujarat and North Gujarat	Resistant to BLB, and wilt and ALS, and tolerant to sucking pests.
RVK 11 (Raj Vijay Kapas 11)	150–160	2.0 (seed); 0.699 (lint)	Tamil Nadu, Andhra Pradesh and Karnataka	Suitable for timely sowing rainfed conditions, average bolls/plant 21.2, tolerant to sucking pests and diseases.
PA 740	150–160	1.36	Andhra Pradesh, Karnataka and Tamil Nadu	Suitable for rainfed, *kharif*, medium to high fertility zones, tolerant to bacterial blight and *Alternaria* leaf spot.

Variety	Duration (days)	Yield (t/ha)	Area of adoption	Salient features
G.Cot.40 (Gujarat Cotton 40)	171–181	2.33	Gujarat, Maharashtra and Madhya Pradesh	Suitable for high-fertility irrigated conditions during *kharif* season, moderately resistant to BLB, *Alternaria* leaf spot and grey mildew.
G.Cot.42 (Gujarat Cotton 42)	165–190	1.74	Gujarat, Maharashtra and Madhya Pradesh	Suitable for high-fertility irrigated conditions during *kharif* season, moderately resistant for grey mildew, compact variety for high density planting system in rainfed areas.
Central Cotton CCH 14-1 (Sunantha)	150	1.69	Andhra Pradesh, Karnataka and Tamil Nadu	Suitable for irrigated tracts during *kharif* season, resistant to bacterial leaf blight, and tobacco streak virus and immune to root rot, tolerant to jassids, white fly, thrips, aphids and stem weevil.

Source of seed and seed treatment

The seed should be collected from approved source. For raising foundation seed crop, nucleus or breeder seed should be obtained. Similarly foundation seeds are needed for raising certified seed crop.

Before sowing the seed should be treated with mancozeb to prevent seed borne diseases. Thereafter, seeds should be rubbed with fine cloth and mixed with cowdung or ash to emerge uniform distribution of seeds.

Seed rate

Seed rate varies according to seed size and spacing. For heavy seeded variety 20-25 kg of seed for hectare (75-90 cm between rows) should be used. But in case of small and medium sized seed it is ranging from 10-15 kg/ha. In case of hybrid seed only 2.5-3 kg seed is sufficient per hectare.

Spacing

For seed production cotton should be always sown in lines. Sowing with the help of deshi plough is a common practice of the farmer, but for maintaining proper spacing and uniform depth use of single row cotton drill is advantageous. Distance from row to row should be 60-75 cm and plant to plant is 45 cm for

American cotton 60-75 cm x 30 cm for deshi variety and 120-150 cm x 60-75 cm for hybrid cotton.

There was a trend an increase in the number of cotton bolls per plant with the subsequent increase in the area for plant, while the number of bolls per hectare showed a trend of decline with the increase in the area per plant. The yield of seed cotton appeared to decline with the successive increase in the area per plant. In hirsutum (variety H14) the maximum yield of seed cotton was obtained when cotton was planted in rows of 75 cm apart at 30 cm plant to plant distance, whereas in arboreum (var R231) the highest yield was recorded in treatment with 75 cm row spacing and 15 cm plant to plant distance. Wider planting caused considerable increase in days to first seed appearance, 50% flowering and boll formation. Row spacing did not exhibit any effect on square maturation period. Number of monopodia and sympodia/plant increased significantly under wider and closer spacing respectively.

Research findings also revealed that the seed cotton yield increased with increase in planting distance. The increase in seed cotton yield in 45 and 70 cm row spacing over 30 cm was 40.3 and 74% respectively. The highest seed cotton yield was associated with 60 cm row spacing. Further, widely spaced crop produced more bolls/plant, boll density/sq.m., boll size, boll weight and seed cotton yield/plant than closely planted one, which might have attributed towards the higher seed cotton yield. Better performance in wider spacing is also confirmed by different research efforts. The higher plant population alone compensated the low yields/plant even through there was less boll weight, nature bolls/plant and seed index in narrower spacing.

Number of bolls picked/plant, weight of seed cotton yield/plant were significantly influenced by plant spacings being maximum under wider spacing. The beneficial effect might be due to less competition for light, aeration, nutrients and moisture which resulted into good growth and yield contributing characters. Thus, the plant population per unit area was the prime factor determining the seed yield. It also indicated that the less seed yield/ plant was compensated by increased number of plants and their seed yield contributing per factors unit area.

5.2.7.3 Nutrient management

Depending on the soil types and yield potential of varieties in different agroclimatic regions, the seed crop should be grown with judicious application of fertilizer.

Well decomposed farm yard manure @ 8-10 t/ha should be incorporated in the cotton field 10-15 days before sowing the crop. In addition to that 50 kg

N, 50 kg P_2O_5 and 50 kg K_2O/ha should be added at the time of final land preparation. Another 50 kg N/ha should be applied in two equal splits in 60 and 90 days after sowing. The seed crop should not be suffered due to deficiency of nutrients. Foliar spray of urea or DAP (2%) at boll development stage is more effective. Research findings also exhibited that the cotton yield per plant increased significantly by nitrogen application upto 100 kg N/ha. Thus 1.57% concentration of nitrogen in petiol was found to be critical, below which there was reduction in yield. Phosphorus fertilization upto 50 kg/ha increased the cotton yield significantly but did not affect the concentration of P in petiol. Hence, phosphorus concentration in leaf petiol did not serve as an index for guiding phosphorus fertilization. Higher amount of nitrogen upto 90 kg/ha increased sympodial branches, boll number and yield of seed cotton.

5.2.7.4 Use of growth regulators

Two foliar sprays of cycocel (first at flower initiation and second at 20 days after first spraying) with 80 ppm concentration increased the bolls/plant, boll weight or N concentration of plant, seed cotton yield and net returns but reduced the boll sheding intensity compared with no spray of cycocel.

5.2.7.5 Interculture operation

Roguing : To get the pure seeds of cotton crop should be rouged at least in three stages i.e. seedling stage, square initiation stage and flowering stage. Firstly, off-type and diseased plants should be uprooted in seedling stage. Second time rouging for off-types and diseased plants should be done at square initiation stage. Lastly, roguing should be done at flowering stage which is the best time to assess cultivar purity.

Weeding: Cotton responds very well to timely hoeing and weeding. Inter cultivation should be done fairly and regularly with either a blade harrow or with a three tyned hoe or with a desiplough so as to keep the plot clean. Inter cultivation not only checks the weed infestation, but also improve the soil aeration and soil moisture conservation. So, intercultivate the cotton field three to four times during the crop period to promote root growth. Hand weeding may be done as and when required. Thinning is desirable for maintaining the optimum plant density to obtain high seed yield. The weeds of cotton field may be controlled by using herbicides i.e. Integrated weed management also can perform well and economically viable for seed production.

5.2.8 Picking

The picking of cotton should be done when the boll is fully mature (boll begin to open). Seed quality also depends on time of picking. Cotton is harvested three or four or sometimes several picking by hand or by mechanical picker as the boll matures over two to three months. Early picking see of cotton performs better germination but better quality of viable seeds may be obtained at the peak of the harvest. The seed collected from the late flush bolls should not be used for seed purpose for the next year.

The following measures should be taken for picking :

(a) Picking of bolls should be done at fully mature stage i.e. when the bolls begin to open.

(b) Picking should be avoided when the bolls are wet due to rain or dew. Start picking with morning hours as soon as the dew is almost dry.

(c) Bolls damaged due to rains or insects should be picked separately (7-10 days ahead of the 1st picking) and should not be used for seed purpose.

(d) The picked cotton should be free from trash, leafbits and other foreign material for reducing the damage during ginning operation through machine.

(e) Moist cotton should not be stored, because moisture content of twelve percent or more may damage the seed.

(f) The picked cotton should be dried completely by repeated sun-drying and then should be stored in a dry place for gining.

If ginning is not done immediately or the picked cotton should be dried completely and stored in a dry place for ginning later on.

5.2.9 Ginning

The ginning of cotton seed occupies an important role for maintaining the quality of seed. This can be done either with roller-gins or with saw-gins. Ginning of cotton seed should be done on one variety gins or gins approved by the seed certification agency. Before ginning the seed crop, machinery should be cleaned to check the contamination. During ginning operation care should be taken to minimize the mechanical damage by reducing the frequency of machining.

The following recommendations are suggested to get the best quality seed after ginning of seed cotton.

(i) Before ginning moisture content of each lot of seed cotton should be determined by moisture meter. It is better to gin only that cotton which contain 6 to 8.5 percent moisture at the feeder apron.

(ii) Limit cleaning equipment to twenty one cylinders only. Control the heat applied to cotton to the lowest possible amount to all proper ginning.

(iii) If possible, use only one saw type limit cleaner may be used.

(iv) The role of ginning should not exceed 4.5 to 5.5 kg lint/saw/hour (on conventional gin stands with 30 cm saws). High capacity gins should follow the recommendation of the manufacturer. All overflow cotton must be returned to the distributor, so as to bypass all precleaning and drying equipment.

Research finding revealed some of the following true ginning aids:

(i) Maintain a loose seed roll at all times.

(ii) Gin saws should be replaced before more than seven percent of the teeth are missing.

(iii) Never mix new and worn saws on the same cylinder.

(iv) Set picker roller as far as possible from hullar ribs without loss of cotton.

(v) Gin saws should be inspected frequently and replaced before they are reduced to more than 1/6" in diameter by wear and sharpening. Maintain correct pitch and tooth shape when sharpening saws.

(vi) Maintain rib spacing at 1/8" at the point where the saw enters between the ribs.

(vii) Brush brittles should mesh into saws not more than the full depth of the tooth.

(viii) Brushes should have a tip speed as recommended by the manufacturer.

(ix) Air flast nozzles should be kept free of all obstruction.

(x) A month pressure of 12 to 14" (30-35 cm) of water pressure should be maintained an air flast nozzles.

Keep safely rocks, clamps and covers in place all times when operating machinery.

5.2.10 Storage

Cotton seed contained high moisture content should not be stored for long period. Seed should be dried with heated air (38-40°C) or combustion gases blown through it. The seed must be cooled before storage. Cotton seed containing upto 12-16% moisture can be stored satisfactorily at 21°C. Above this temperature, it is unsafe to store cottonseed in quality when the seed moisture is above 11-12%. Heat is generated in the store cottonseed due to the action of respiratory enzymes and is not rapidly dissipated owing to the insulating effect of cotton fuzz unless adequate precautions are taken to prevent a rise in temperature. The glycerids present in the seed may be metabolized to yield free faty acids, further deteriorating the seed quality. Storage houses, therefore, must be well ventilated or provided with forced air circulation to prevent a rise in storage temperature.

5.2.11 Seed Treatment

A spray of 4-5 kg of a mixture of propylene glycol dipropionate and bischloromethylene perton of cotton seed prior to its storage has been recommended to prevent heat development and seed damage. Treatment with Nacconol (sodium alkylaryl sulfonate) also gives satisfactory results.

5.2.12 Seed Yield

Average seed yields of cotton seed vary from 300 to 600 kg/ha.

6

Seed Production of Potato

Potato is an important crop of the world and grown in almost all the states in India. Apart from India, potato is grown in more than 100 countries but the intensity of cultivation varies considerably. For example in the Netherlands 19.6% of the total arable land is cultivated with potatoes. The figures for Germany, Peru and USA are 10.1%, 5.6% and less than 1% respectively. India ranks fourth in area (1.29 m ha) and third in production (25.23 m ton) in the world. The productivity of potato depends mainly on the quality and quantity of planting materials and the agronomic management practices adopted. Most of developing countries including India experience release of high yielding varieties, development of adequate agronomic and pest control measures but the gap between supply and demand of good quality adequate quantity seed has to be the major constraint in increasing its productivity. The seed production in India is mainly confined to the North-Western part both in the hills and the plains, because of favourable weather and low incidence of aphids (vectors responsible for viral diseases) in this regions. The seed produced in our country is much less than the requirement. Thus, against an annual requirement of about 6.5 thousand tonnes of breeder's seed required to produce enough certified seed after three years of multiplication (considering that the farmers' replace their seed stock only once in three years), only about 2 thousand tones are being produced presently. The gap in the demand and supply of good quality seed is generally filled with quality seed available in the market.

To overcome the problem of shortage of breeders' seed of improved potato varieties a number of steps have been taken. These include :

(a) Involvement of State Agricultural Universities in breeders' seed production : Six Agricultural Universities namely, HPKVV, (Palampur), CSAUAT, (Kanpur), BCKV, (Mohanpur), GBPUAT, (Pantnagar), RAU, (Pusa) and HAU, (Hisar) have been identified in the first phase for participation in the breeders' seed production to meet the requirement of respective states. Breeders' seed is produced by these universities under the supervision of a team of scientists from CPRI, Shimla which is constituted to formulate, oversee, supervise and monitor the programme and to fix the targets of breeders' seed production;

(b) An integration of micro-tuber production through tissue culture technique for fast multiplication of disease free stocks. The technique for production of micro-tubers has been standardized and the micro-tubers are grown in controlled conditions in nursery beds and subsequently in field to produce breeders' seed stock. This technique reduces field propagation of indexed tubers to two years thereby ensuring a faster multiplication and better quality of breeders' seed;

(c) An improved virus detection ELISA (Enzyme linked Immuno Sorbant Assay) technique has been introduced in the seed production programme for detection of viruses X, S and Y. The virus detection with this technique is nearly 20 times more efficient than that of serological testing;

(d) Identification of new areas for breeder seed production e.g., Hisar, Kanpur, Deesa where quality seed can be produced under low aphid population. The exploitation of these areas for seed production will further increase the availability of seed potatoes in the country.

6.1 Botany

Potato (*Solanum tuberosum*) is essentially a plant of temperate region with long day condition. In India 90% of the crop is cultivated during winter in plains under short day conditions. Short winter days (10-11 hrs sunshine) influence the growth in a variety of ways, but the main effects are suppression of vegetative growth and flowering. The stems may be light to dark green, erect and hard, firm and spreading or weak and flexible. Thickness of the stem produces significant role in tuber production. The thick stemmed plants usually yield large size tubers whereas these stemmed plants produce small sized tubers (Sharma, 2002). Underground stem known as stolon arises laterally as auxillary buds on the underground portion of the stem and grows along the horizontal plane in the ground but sometimes emerge above the ground. The underground modified stems i.e. tuber is an enlargement of the terminal end of the stolon. The leaf is compound and pinnate with several pairs of leaflets arising in succession, along the rachis and an odd, large terminal leaflets. The potato flower is a complete flower as it contains all the four essential parts. The calyx is composed of green sepals united to form funnel shaped structure and persistent type. Wheel like 'rotate' type well developed corolla has 5 petals which are light-blue to deep blue and from faint blue-purple to deep blue purple (lavender). It may be often white. It is distinguishing character of varieties. The androcium retains five light yellow to deep orange colour anthers. Some varieties do not contain any pollen and are sterile but varieties like Kufri Safed are highly fertile so fruit setting is very common (Sharma, 2002). The ovary is well developed. After fertilization ovary develops into small (2-4 cm in diameter), round, green berry (fruit). Each berry contains numerous seed embodied in the soft flesh pulp within the very. A tuber (modified under ground stem) has 2 ends *viz.* rose or apical and heel end. The apical end represents the growing point and lies towards the outside. Heel end attached to the stolon. The each eye bears may buds. At the rose end more eyes are present than that of heel end. Apical eye sprouts first and influence the sprout and performance of other eyes. The shape of the tuber depends normally on variety; normally round to cylindrical in shape (Agarwal 1999).

6.1.1 Flowering

For sexual reproduction leading to true potato seed (TPS) flowering of the parental lines is essential. Although thermo-photoperiod does not play a role in flower induction, yet the bud development and blooming are dependent on long day length (> 14 hrs) under cooler climate. In general, andigena group bloom profusely even under equinox conditions, the tuberosum group needs long photoperiod (>14 hrs. effective daylight). In highland tropics and sub -tropics, where potato is grown under short day or equinox conditions, the day-length is extended to > 14 hrs. using high pressure sodium vapour lamps providing minimum required light intensity of 50 lux at the plant level.

6.1.2 Breeding system

Most potato species (wild and cultivated) are diploid (73%) and a few are triploid (4%), tetraploid (115%), pentaploid (2%) or hexaploid (6%). The majority of the diploid (2X) potato species are outbreeders due to gametophytic S. allele incompatibility system. They are variable in nature and show marked inbreeding depression. On the otherhand, the tetraploid species are highly self-fertile, and the disomic polyploids (4x and 6x) are primarily self pollinating. So the polyploid are less variable and tolerate inbreeding. However, the Tuberosum group (*S. tuberosum* ssp. *tuberosum*), the potato of commerce show some outbreeding nature due to protogy and stigma protrusion and at the same time inbreeding depression.

6.1.3 Crossability

Crossability between diploid species is good, suggesting basic genomic similarity between them. The F1 progenies of these crosses are fertile. However, some groups of diploid species are not crossable. Subspecies of *S. tuberosum*, the subsp. *tuberosum* and subsp. *andigena* are intercrossable. In most of the cases there is little genomic differentiation within the tuber bearing *Solanum* species. Mostly 'A' genome is found,. For this reason, regular hybridization between the species are possible.

Interspecific hybridization is feasible among the following species.

S. tuberosum ssp. *andigena* (4X), *S. tuberosum* ssp. *tubersum* (4X) *S. stoloniferum* (4X), *S. acaule* (4X), *S. stenotomum* (2X), *S. phurija* (2X), *S. chacoense* (2X), *S. vernei* (2X), *S. bulbocastanum* (3X), *S. juzepczukii* (5 X), *S. curtilobum* (5 X).

Such success in crossability among the species having various ploidy level suggest that chromosome variation does not pose much barrier for interspecific hybridizatgion. Rather, taxonomic relationship of the species largely determine the crossability. However, odd ploidy, endosperm balance and allopolyploidy (different genomic constitution) are some of the important crossability barriers those are difficult to overcome.

6.2 TPS Technology

True potato seed (TPS) technology is a radical departure from the traditional method of production. Which is based on the use of seed tubers. One of the news areas of potato research connected with development of this novel technology at the International Potato Centre (CIP) in Peru is researchd on the production of high quality TPS. True potato seed, a sexually produced propagule and popularly known as TPS, was well known to the farmers in Andes (South America) since the time immemorial. The farmers in this part of the world have been using this entity for rejuvenating the planting material that was either degenerated or destroyed by disease epidemics. The potential of TPS as a low cost alternative for potato cultivation was realized during early fifties in India and China. Utilization of TPS for commercial potato production raises the status of this propagules as an economic entity similar to the certified high quality seed tuber. As such the technology to be developed for TPS production on commercial scale assumes a different dimension. The major advantages of this technology over the traditional seed tuber technology are as follows (CPRI, 1999).

1. Unlike the seed potato production which is confined to northern India only, the TPS can be produced in all potato growing regions.
2. The crop raised through TPS is almost disease free as most of the disease get filtered out during TPS production.

3. The TPS being very small, can be stored and transported easily, whereas the seed tubers are bulky, hence the storage and transport are expensive.

4. TPS provides a low cost potato production technology where only about 50 g TPS is required for sowing in about 375 m^2 area for producing seedling tubers enough for planting one hectare next year. About 150 g TPS/ha is required if the commercial crop is to be raised in the first year itself by transplanting seedlings in field.

The TPS technology involves two major steps

(1) Production of hybrid seeds

(2) Its use as planting material for raising commercial crop. TPS hybrids namely TPS C-3 and HPS 1/13 have been recommended for Assam, Bihar, Gujarat, Haryana, Meghalaya, Karnataka, Western UP and Gujarat, Satputra plateau of M.P., respectively. The hybrid seed of TPS C-3 is produced using the culture JTH/C-107 as female and JEX/A 680-16 as male parent, and the TPS of HPS 1/13 is produced using the culture of MF1 as female and TPS-13 as male parent.

For commercial use TPS can be utilized directly for raising crop by first raising the seedlings in nursery beds and then transplanting them in fields or by first producing seedling tubers in nursery beds/fields which are subsequently used for raising the commercial crop next year. The effectiveness of the methodologies depend on the area of its use, environment and resource availability.

6.2.1 Production of Hybrid TPS

The hybrid TPS can be produced both in the hills and plains. In hills, the crop is grown in long summer days and flowers naturally and hybridization for seed production can be achieved easily. Whereas in plains the crop for seed production is grown in winters so additional artificial light has to be arranged to induce flowering in the crop. The crop needs moderate temperature for good flowering and seed setting. In northern plains, where the winters are severe, the parental lines need to be planted either about 15-20 days before optimum time of planting of the crop, or the planting is delayed till the second week of November so that the flowering time does not coincide with the severe winter period. Delayed planting of parental lines should not be done in first prone area.

6.2.2 Selection of Parents

In the hybrid TPS production programme parents are generally selected from the stocks of *S. tuberosum* spp. *tuberosum* and *S. tuberosum* ssp. *andigena*.

Generally *andigena* lines are utilized as male parental lines, because tuberosum lines/cultivars produce little or no viable pollen. The parental lines are selected through the analyses of combining ability and performance of the resultant hybrids. The selected parental materials are maintained and multiplied inside the aphid proof net house to avoid deposition vector transmitted virus to the parental lines. Seed tubers of the parental materials are kept in cold storage.

6.2.3 Characteristics of TPS parental line

Male parents having indeterminate growth, copious production of fertile pollen, stable/durable resistance to late blight and higher combining abilities for desired horticultural traits are selected from group andigna. Genotypes with tuberosum background, late blight resistance determinate growth, low pollen fertility or stable tetrad sterility and high combining abilities for desired horticultural traits are used as female parents.

6.2.4 Maintenance of parental lines

This is highly technical, skillful and research-oriented part in the whole process of hybrid TPS production. Maintenance of TPS parental lines in disease-free condition needs to have well developed infrastructure facilities for sustainable supply of pathogen-free seed tubers every time needed for planting in crossing block.

6.2.5 Land preparation

The field is to be prepared as for ware potato crop, by adding farm yard manure (FYM) @ 20 t/ha, ploughing, planking and pulverizing clods to bring soil to a fine tilth.

6.2.6 Planting crossing block

Staggard planting of parental lines facilities better management of pollination in the crossing block for hybrid TPS production. Male parents are planted about a week earlier than the female lines to have pollen available before the female lines start blooming. The area required for planting male block is generally kept about ¼ of the female block, depending on pollen producing ability of the male parent. In the blocks, furrows at 60 cm distance are drawn. Normal dose of fertilizers as recommended for ware crop are applied. Tubers of 30-60 g are planted in furrows at 20 cm intra-row spacing. The seed tubers of the male and female parents are planted in the ratio of 1 male : 5 male. In the female block, beds of three rows each are prepared. For this, 3 rows at 50 cm inter-row distance (in the hills) or at 45 cm inter-row distance (in the plains) leaving 80 cm walking space between two adjacent beds are made. In these blocks tubers are planted at 15 cm distance. After the germination, trim the plants in female block to retain a single stem per plant.

6.2.7 Trimming flowers

In female parents, flower buds are trimmed to six flowers per bunch/stem to obtain higher fraction of bold seeds as the essential constituent of high quality TPS. Plants retaining 6-8 large size buds per bunch is essential for quality production. Very small buds, old flowers and berries if any, should be removed from the bunches prepared for pollination next day.

6.2.8 Arrangement of light

After the emergence of the plants above ground level, lighting arrangement to provide 4-5 hours extra lumination everyday after sunset with 30-40 lux intensity is to be made to induce flowering. One sodium vapour lamp of 150 W per 100 sq.m. is sufficient to induce the flowering effectively. Flower buds starts to appear 35-40 days after planting.

6.2.9 Flower collection

Flowers, from male parents, are collected just before the anther dehiscence in the evening preceeding the day of pollination to obtain maximum pollen. Only just opened flowers with anthers that are about to shed pollen or the large size buds, which would open next day, should be collected.

6.2.10 Pollen extraction and storage

Flowering of male parents occur earlier than that of female as the male parents planted a week earlier than female lines. Collected flowers from male parents are spread over thinly on the sheet of blotting paper to make the flower dry. After that, the anthers are separated from the flowers and spread over the sheet of blotting paper for drying by circulation of air for 4-5 hours. After initial drying over the blotting paper, the surface disinfestations of seeds. The seeds are washed thoroughly to ensure complete removal of the chemicals. Then seeds are spread in a thin layer on muslin cloth stretched over a wooden frame and kept the frames in a well ventilated room preferably under fan for 24 hours, or in shade (in the plains) or in mild sun (in the hills) for drying for three days. It can also be safely dried in any type of low moisture environment i.e. forced air ovens, fan etc. Temperatures above 30°C should be avoided during the initial seed drying period (upto 12 hrs.). Thereafter seeds can be safely dried under temperature not exceeding 40°C ± 5°C. With any method used to dry, the newly harvested seed should be dried to a level of about 4% seed moisture content.

6.2.11 Studies on germination of seeds

The seeds extracted from different grades of berries were subjected to germination studies with 100 seeds each from different grades.

Table 6.1: Mean seed germination value

Berry grade	Mean weight/berry	Germination %
Large	8.5	61.6
Medium	5.5	70.3
Small	3.0	63.3

The results showed no significant differences among the seeds obtained from three grades of berries (Dissanayake and Devasabai, 1988).

6.2.12 Processing TPS for high quality

The following steps are to be followed to obtain high quality TPS with high sowing value and high viability.

6.2.12.1 Winnowing

The dried seeds (TPS) are cleaned properly. To remove the empty/brown/aborted seeds are chaff, a seed blower is operated at the desired airflow. By regulating the air flow, the light seeds are also separated to obtain heavier seeds.

6.2.12.2 After ripening of TPS

Hybrid TPS has a dormancy period of 4-6 months depending on the parents involved. The TPS also emerge very poorly under high temperature conditioned (above 25°C). This is the most crucial and important aspect for attaining a high sowing values of TPS. After drying a desired moisture level, the seed should be after ripened at 30°C for different durations depending on the TPS family. This helps in the release of dormancy syndrome as well as setting all the seeds at the same physiological stage. This step is required for uniform seedling emergence and growth in the nursery beds, and better transplanting efficiency in the field.

Anthers are placed in the fine mesh nylon net bag and kept inside a desicator for proper dehydration. The desiccation containing the anthers in the nylon bags is kept inside a refrigerator overnight at a temperature of about 8-9°C. In the next morning, the anther containing nylon bags are taken out from the desicator to a dry and closed chamber. The pollen grains are then extracted in a petridish by mild shaking the anthers through a fine mesh nylon net sieve. A vibrator can be employed for such extraction of pollen grains. If through extraction of pollen grains from the anthers is not possible in the morning, those anthers are again dehydrated by keeping them inside a desicator in the same manner. The second stage of pollen extraction is done in the next morning.

The pollen after extraction are dried sufficiently in the room temperature and stored in small (5-6 mm) air tight vials. The vials are kept inside a refrigerator at a temperature of 8-9°C. Pollen remain viable inside the refrigerator for about 14 days and up to 1 month at 2.5°C temperature.

6.2.12.3 Pollination method

Higher pollination efficiency is achieved by dipping receptive stigma into the pollen grains collected in shallow vials. This method is superior in comparison to other methods such as a brush or the pair of forceps used to transfer pollen grains to the stigma in female parents. First pollination is to be performed between 8-10 AM and repollinate the flowers at the same time next day. The processes of pollination should be continued till the flowering period of the crop is over. Usually there are two flushes of flowers during the crop period. Pollination should be done abundantly otherwise hybrid seeds per berry will be reduced. Left over pollen is stored in small vials; kept the vials in a desicator and placed the desicator in refrigering for at 6-10°C for use next day, if necessary. But it is always advisable to use fresh pollen for pollination every day.

After successful pollination, the stigma had turn brown and the flower slightly droop. The fruit i.e. berry is set within a week of pollination. If necessary then support is to be given to the stems of female plants so that the flower bunches and the berries do not touch the ground. After the formation of berries, the berry bunches should be covered with thin muslin cloth bag of about 8 cm x 12 cm size.

6.2.12.4 On-plant maturation and harvesting of berries

For producing vigorous or high quality TPS, higher doses of nitrogen are applied in splits at every 15 days interval to delay mother plant senescence and to extend on-plant maturation period of berries. The matured and yellowish green berries are collected 50-55 days after pollination ensure better quality TPS.

6.2.13 Post-Harvest Ripening of Berries

Berries are kept in room temperature for 10-15 days which makes them soft for easier maceration.

6.2.14 Extraction of Seed

A manually operated reverse screw juice extractor is used for maceration of the berries (100-150 kg a day). The debris of the crushed berries is strained through the sieve nylon net bag of 3.13 mm pore size or 20 mesh. This facilities removal of bigger pieces of fruit tissues simultaneously allowing TPS to pass through into water in the containers. Water is run at high pressure, over the TPS collected in the containers, to remove mucilage, bits of fruit tissue and to hasten the process of obtaining clean seeds. Then seed treatment is to be done for surface disinfection. Seeds are treated with 10% hydrochloric acid (HCl) at continuous stirring for 20 minutes. After thorough rinsing with water followed by 0.5% sodium hypochlorite treatment (10 minutes) removes mucilage and helps in obtaining clean seeds.

6.2.15 Dormancy Release and Vigour Test

TPS should be analysed for determining dormancy and viability according to the protocols developed at CIP Head Quarter in Lima, Peru. TPS are checked for germination at 27°± 1°C in the incubator. More than 50% germination obtained in 7 days, indicates high sowing value, good seed vigor and release of dormancy in TPS lot. Dormant or non-viable TPS shows low germination performance. However, after 7 days at 27°C, the seeds when shifted to 18°C, if start germinating, indicates incomplete release of dormancy. But if they fail to germinate then the TPS lots is considered as non-viable.

6.2.16 Packing and Storage of TPS

After washing, cooling in shade, the seeds are packed in moisture proof packets like, plastic packet, alluminium foil lined packets etc. The seed packets are stored in cool and dry place. Sealed packets are stored in dessicator kept in refrigerator at 6-10°C or even at room temperature during the winter.

In general, the total weight loss and percent rottage were higher in TPS population in comparison to the commercial varieties at ordinary room temperature (Table 6.2). However, an experiment by Singh (1988) showed in TPS C1, C3 and C16 weight loss and rottage were comparable with K. sindhuri and K. jyoti. Thus, in this respect also, TPS population were equally good.

Super quality TPS is non dormant seed > 1/18th size with high quality and high sowing value giving uniform germination/emergence, vigorous seedlings transplantable in 21-25 days, having higher transplanting efficiency and tuber yield (CIP, 2000).

Table 6.2: Rottage and weight loss during storage at ordinary temperature in TPS population

Genotypes		Loss		Sprouting behaviour	
		Loss in weight (%)	Rottage (%)	Dormant (%)	Sprouted (%)
TPS	C 1	35.0	26.6	85.7	14.3
TPS	C 3	44.5	38.2	57.8	42.2
TPS	5	94.0	93.3	77.7	22.2
TPS	13	86.5	82.7	21.0	79.0
TPS	16	43.5	36.1	72.5	27.5
TPS	17	65.0	64.0	39.1	60.9
TPS	19	90.5	84.8	57.8	42.2
K. jyoti	55.0	46.9	67.4	32.6	67.4
K. sindhuri	62.0	57.1	59.0	41.0	59.0

6.2.17 Production of Seedling Tubers from True Potato Seed

Grewal and Kang (1990) reported that in hills flowers form and seed are produced . These seeds are supplying in the plains. Seedling tubers in nursery beds can be produced by two methods.

Methods-I: In this method, the seedling are raised in primary nursery beds and subsequent transplanted in secondary (production) nursery beds for tuberization. About 40 g TPS and 30 m^2 nursery bed area required for producing enough seedling tubers for planting one hectare area in the next year.

Method-II: In this method, 2-3 seeds at 10 cm intra-row and 10 cm inter-row are sown. Depth of sowing is 0.5 cm. The seedlings are not transferred to another seed bed.

6.3 Seed Potato Production

The sprouted tuber is planted for production of potato. The productivity or yield primarily depends on the quality of seeds. Quality seed implies that the seed tubers should be (i) pure and free from varietal mixture, (ii) free from diseases and pests, (iii) viable or physiologically sound, capable of sprouting readily, and (iv) of optimum size depending on the variety and purpose for which potatoes are grown. To achieve good quality of seed, a sound and scientific seed production programme has to be developed.

Seed potato production primarily confound to (i) summer crops in high hills in north western hills of Himachal Pradesh, (ii) Autumn crop in the plains. The seed produced in these area is supplied to other part of the country for raising commercial crop (ware crop). Besides these, the north-eastern hill regions and Nilgiri hills grow their own seed potatoes for inter-regional distribution. Midium hills and low hills are generally not suitable for seed production because the conditions in these areas are conducive to the spread of viral diseases and bacterial brown-rot disease. Similarly, plateau areas are also not suitable for seed production because of the early appearance of aphids causing transmission of viral diseases (Sharma, 2002).

Most important achievement in potato seed production is the development of seed plot technique based on aphid surveys. This indicates the weakness of traditional seed production programme and identification of low aphid periods in the plains. The following steps are to be taken for viable seed production in plains.

6.3.1 Agronomic Management

6.3.1.1 Land preparation

The field for seed production is ploughed during the summer months of April-June. Then the field is kept open or bare and one or two turnings to the soil during hot summer days is given to reduce the incidence of soil-borne disease and pests, also to control perennial weeds (Agarwal, 1999).

Potato does not grow well under wet and ill-drained soil. So, well drained sandy loam and loamy soil are most suitable for potato cultivation. Field should be thoroughly ploughed with mouldboard plough. After green manuring, preparation of land is to be done. Land should be leveled and there should be good provision for good drainage.

6.3.1.2 Green manuring

Green manuring crops like Dhaincha (*Sesbania aculeata*) or sunnhemp (*Crotolaria juncea*) may be sown in the second or third week of June, just before the onset of monsoon. The green manure crop is buried in later half of August with mouldboard plough. Generally, selection of green manure is done based on land situation; sunnhemp is planted in upland, whereas Dhaincha is selected for lowland. Green manuring improves the physical condition as well as nutrient status of the soil (water holding capacity and increase the nutrient availability to the potato crop).

6.3.1.3 Planting variables

Only foundation or certified seed should be used for multiplication. These seeds should be procured from the state agriculture horticulture department or the National or State Seed Corporation. The seed stock should be replaced after every 3-4 years.

The pre-sprouting seed weighing about 40 g each and having multiple sprouts is used to produce seed tubers. Pre-sprouting is of advantage in seed crop and in multiple cropping as it reduce the time for crop emergence. Cut tubers should not be used for raising the seed crop.

In the high hills, seed from the stores should be removed in the gunny bags by the time dormancy is over and be kept in different layers in wooden baskets or on the floor in a well ventilated store exposed to diffused light in a low moisture condition for proper sprouting. In plain, seed tubers are to be removed in September – October (at least 7-10 days before planting). Seed bag should be kept in a pre-cooling chamber of the cold storage for at least 24 hours otherwise the seed becomes moist immediately after it is removed from cold storage because of condensation of water vapours on the tuber surface. It may result in rottage due to immediate exposure to high temperature. For pre-sprouting the seed tubers are kept in an environment consisting mild temperature of about 20°C, moderate high humidity of 60-80% and diffused light.

The sowing/planting of seed crop depends on prevailing temperature, agroclimatic situation and crop rotation, etc. Optimum time for planting in plain is during the last-week of October to first-week of November. Earlier planting results lanky plants whereas delayed planting causes heavy infestation of aphids.

Planting Techniques

(i) Keep the seed in furrows already made for the application of fertilizers. When planting is done by tractor, spacing between the rows is kept at 60 cm and between the tubers at 20 cm. Cover the tubers with soil after planting using a ridger or spade.

(ii) Potatoes are also planted by dibbling on ridges. For this also open furrows 60 cm apart with narrow spade or tractor driven marker in east-west direction and keep the soil on the southern side forming the ridges. Place the fertilizer mixture on the northern side of ridge and cover it with 3-4 cm of soil. Dibble the tubers on the ridges at a distance of 20 cm and depth of 5-6 cm. An alternate method is to use a tractor driven fertilizer drill-cum-potato planter, or a fertilizer drill-cum-marker for making furrows. By using the instrument, the number of field operations can be reduced.

Seed rate depends upon tuber size. Twenty five to thirty quintal of seed potato per hectare will be sufficient if the usual sized tuber (4 to 6 cm diameter) are used.

6.3.1.4 Nutrient management

(i) If green manuring has not been done, broadcast well decomposed farm yard manure (FYM) or compost @ 250-300 q/ha at the time of land preparation.

(ii) The optimum dose of inorganic fertilizer depends on the soil type, soil fertility, crop sequence and related factors. The seed crop should be supplied with 100 kg nitrogen (4.0 q calcium ammonium nitrate), 60 kg phosphate (3.75 q single super phosphate) and 80 kg potash (1.33 q muriate of potash) at the time of planting and 50 kg nitrogen (2.0 q calcium ammonium nitrate) per hectare at the time of earthing up. If FYM or compost has been applied as mentioned above, the doses of potash and superphosphate can be reduced by half. Excess application of nitrogen should be avoided as it will be difficult to rogue out diseased plants from such plots and it will result in more number of large sized tubers with consequent reduction in the number of seed sized tubers. Apply fertilizer in furrows, then cover the furrows partially with soil so that the tubers do not come in direct contact with fertilizer (CPRI, 1987).

6.3.1.5 Interculture

After 25-30 days of planting, when the plants are 8-10 cm high, remove the weeds using a tractor-driven spring- tyne cultivator between the rows or by a khurpi. Weeds in seed crop are, however, better controlled by spraying Paraquat dichloride @ 2.5 l/ha dissolved in 1000 litres water. The weedicide spray is given when the plant emergence is about 5%. Then apply the remaining dose of nitrogen. Mix it with a spring-tyne cultivator and earth up the soil to make a thick ridge with the help of a narrow spade or tractor driven potato ridger. Care should be taken that the plants are not disturbed during cultural operations (CPRI, 1987).

6.3.1.6 Irrigation

One irrigation before planting is advantageous for ensuring uniform germination. If pre-irrigation has not been given, then the first irrigation should be given soon after planting. This post-planting irrigation should be light and subsequent should be given at 6-10 days intervals in sandy loam soil and at 10-12 days in heavy soils. Normally, the soil should remain moist but not become too wet. Care should be taken to see that the ridges are not submerged under water in any case. Stop irrigation 7-10 days before haulm cutting (CPRI, 1987).

6.3.1.7 Roguing

During crop season, examine the seed plots at least thrice to remove off-types and diseased plants, e.g. those showing mottling, mosaic, veinal necrosis, crinkling, rolling of leaves, marginal flavescence and purple top roll. First roguing is done 20-30 days after planting and immediately before earthing up. Second roguing should be done 45-50 days after planting. Care should be taken to ensure that along with the diseased and off type plants, all the tubers (that means the underground portions) are also removed. Third roguing is done 3-4 days before haulm cutting. Collect diseased plants in a bag and bury in a pit away from the field (CPRI, 1987).

Roguing is one of the major operations for high quality crop of seed potato. Roguing is to be done thrice (at least) to remove off-types and diseased plants e.g. those showing mottling, mosaic, veinal necrosis, crinkling, rolling of leaves, marginal necrosis, crinkling, rolling of leaves, marginal flavescence and purple top roll. The roguing is to be done at the following three stages.

First roguing

First roguing is done 20-30 days after planting and immediately before earthing up. In first roguing, all virus affected plants and plants of other varieties (identified from morphological character) should be removed from the seed production plot.

Second roguing

Second roguing is done between 45-55 days after planting. In this operation care should be taken to ensure that along with diseased and off-type plants, all the tubers are also removed.

Third roguing

This final roguing is done 3-4 days before haulm cutting. At this stage, all the virus affected plants and off-type plants, if any, along with their tubers have to be very carefully removed, so that no such plants are left in the fields.

6.3.1.8 Haulm killing

It is an important practice adopted as a precautionary measure to avoid chances of viral disease transmission through the vectors like aphids. It is also effective in case the blight appears.

It restricts the chance of tuber infection by spores. Irrigation should be stopped 7-10 days before haulm killing. Haulm is to be cut when 20 aphids per 100 compound leaves are observed. Normally it is done during mid-January. Plants are cut at ground level. It is to be ensured that all regrowths appearing on the stumps after dehaulming are promptly removed, as tender and succulent leaves attract the aphid vectors. Gramaxone @2 litre/ha would prevent regrowth and has also been effective in controlling weed. Even two sprays at 4-7 days interval are taken to get complete killing of haulms.

6.3.2 Harvesting and Curing

(a) Time of harvest: The crop is ready to harvest 10-15 days after haulm killing when the skin of tubers become firm. The harvesting may extent upto 15 February and never be delayed beyond 3rd week of February; this causes in tuber rot due to soft rot and charcoal rot. Even early premature harvesting should not be recommended as the skin of the tubers are soft and easily peeled off, such tubers can not withstand long transplantation and storage.

(b) Soil condition: For harvesting clean and whole tuber the soil moisture condition should be optimum.

(c) Method of harvest: Harvesting is better done by tractor driven or bullock drawn potato digger, as it causes less tuber damage than when done by Khurpi. Every effort should be made to avoid cuts, buries etc.

(d) Curing: After harvest, the potatoes should be kept in heap in a cool pace (preferably around 15-20°C) for about 10-15 days for curing the skin. The heaps can be 1-1.5 m height and 3-3.5 m wide at the base. In Indo-Gangetic plains, rice straw is used to cover the heaps which protects the heaps from direct sunlight and maintain equable temperature. If the crop is harvested under high relative humidity (as in hill) before going for curing, seed potato is spreaded in a thin layer to reduce moisture content.

6.3.3 Grading

Grading of potato tuber is an important operation depending on the shape, colour, depth of eyes, weight, disease infestation, etc. All the off-types, cut damage and rotten tuber should be removed. Several disease may affect the seed tuber quality. These diseases can be recognized by observing few distinct features. These are described below.

Table 6.3: Disease symptom on seed potato tuber

Disease	Symptom
Late blight	Sunken purple spots on the surface of the tuber, flesh has brown necrosis
Charcoal rot	Near eyes and lenticels black patches appear
Black scruff	Chocolate colour sclerotia on the tubers
Dry rot	Brown patches enlarge and sunken ultimately skin wrinkled to form concentric ring

If these symptoms have been observed the affected seed potato tubers should be removed. Finally grade the tubers according to their size preferably in four classes *viz.,* small (below 25 g), medium (25-50 g), large (50-75 g) and extra large (above 75 g).

6.3.4 Treatment of Seed Stocks

Some pathogen infect the tubers but do not show any visual symptom. Finally, these diseases transfer to next crop, so seed treatment against seed-borne disease with 3% boric acid for 20-30 minutes provides effective result.

After grading, the seed tubers should be washed properly. After washing, to provide protection against infectious disease tubers may be dipped in 1% bleaching powder solution. Then after properly cleaning the seed tubers 0.5% organomercurial compound solution is used for 30 minutes for protecting against surface borne disease. After completion of these treatments seed tubers must be dried in shade and packed in gunny bags. The care should be taken that the seed tubers must not be used in ware purpose.

6.3.5 Seed Storage

Seed tubers must be taken for storage before 15th March as the temperature beyond mid-March gradually increases. The bags should be checked properly for distinguishing them from ware tubers. Before going for storage, the bags should be kept in pre-cooling chamber. The temperature inside the cold store, should be maintained at 2-4°C with 75-80% relative humidity during storage. The chamber must be inspected periodically. There should be arrangement of free air circulation.

Table 6.4: Insect management of potato seed production

	Insect pests	Symptoms	Control
(i)	Aphid	Sucking the sap causing poor growth of the plant, peak period December to February	(1) Phorate 10G @ 1.5 kg/ha at the time of first earthing up (2) Rogor or Metasystox at 625 ml per ha.
(ii)	Leaf earthing caterpillar	Young leaves attack first, chewed the leaves	(1) Endosulfan @ 1.5 ml/1 of water or with carbaryl @ 2.5 kg/ha
(iii)	Cut worm	Damage young plants by cutting them above the base near the soil in the night	(1) Dusting endosulfan, methyl parathion, quinalphos @ 30 kg/ha before sowing and at the time of earthing up. (2) Spraying with chlorpyriphos @ 2.51/ha in 1000 litres water
Disease			
(a)	Late blight [*Phytophthora infestans* (Mont.) de Bary]	Brown dead spots or extended dead areas appear at the tip of margins of the leaves and spread downward or inward	0.2% solution of Mancozeb may be sprayed at 10-15 days interval.
(b)	Early blight (*Altenaria solani* (Ell. And Martin) Jones and Grout]	In the necrotic tissues concentric ridges develop to produce a target board effect, lowest leaves are attacked first and the disease progress upwards.	0.2% Dithane M-45 (Mancozeb) or Dithane Z-78 (Zeneb) sprayed at 10-21 days interval.
(c)	Black scruf (*Rhizoctonia solani* Kuehn)	Tuber infection appears as black crust on the skin due to formation of sclerotia of the fungus, these are normally skin deep.	Seed treatment with 0.25% suspension of an organo mercurial fungicide containing 6% mercury

Table 6.5: Important Potato Varieties

Name of the variety	Duration (days)	Yield (t/ha)	Cultivated area	Characteristics
Early duration crop:				
Kufri Chandramukhi	80-90	25	Indo gangetie plains Madhya Pradesh Plateau region	Medium tall, vigorous having open look with lighter purple flowers
Kufri Navtal (G-2524)	75-80	15-20	Most suitable for Kumaon hills	Tuber flesh is light yellow and texture is waxy. Good carotene (Vitamin A) content and very good cooking quality. Moderate resistant to late blight at field level.
Kufri Lavkar	75-90	25	Northern plains Tropical Plateau region	Tuber flesh is white, posses slow degeneration, partial resistance to blight and immunity to wort disease.
Kufri Alankar	75 to 130	20-25	Northern plains	Tubers large, white with shallow eyes. Keeping quality is poor, thus does not store well in cold storage.
Kufri Jyoti	80-100	20	North Eastern Hills and plains, U.P., Kumaon Hills Nilgiri Hills	Tubers are susceptible to cracking under long duration of plants. Not suitable for long distance transportation.
Mid-duration crop:				
Kufri Navjot	125-130	20	Hill and Plains of Northern India	Tuber flesh is dull white and texture is waxy. Moderate degree of field resistance to late blight. Somewhat resistance to late blight.
Kufri Chamatkar	110-130	25-30	U.P., Punjab, Haryana and Rajasthan	With white medium deep eyes and yellow pale flesh. Resistant to early blight and susceptible to mosaic and leaf curl.
Mid-duration crop:				
Kufri Sheetman	110-130 days	25 t/ha	Northern India	Tubers oval, medium to large sized with shallow eyes, white and texture floury on cooking.

Name of the variety	Duration (days)	Yield (t/ha)	Cultivated area	Characteristics
Late-duration crop:				
Kufri Deva	120-125	30-40	All the plains of India	Medium sized oval tuber, pale yellow frost tolerance with good keeping quality at room temperature as well as cold storage. Susceptible to late blight and resistance to visible mosaic and stress.
Kufri Sindhuri	120-140	30-35	Plains in Northern plains	Tall, erect plant with blue purple and abundant flowers. The tuber flesh is pale yellow and texture is waxy.
For chips				
Kufri Chipsona 1	100	25-30	North Eastern plains and Hill	Medium tall, semi erect, medium compact plant with white flowers. Tubers with low reducing sugars, low phenols and 2-4% higher dry matter than existing varieties.

Recently released varieties / hybrids of potato

Variety	Duration (days)	Yield (t/ha)	Cultivated area	Characteristics
Early duration				
Kufri Ashoka	70 – 80	35-40	Indo-Gangetic plains of Bihar, Haryana, Punjab, Uttar Pradesh and West Bengal	Medium tall, erect, medium compact and vigorous plants with green foliage and light red – purple coloured flower. Tubers are white, large, oval to long, smooth skin, fleet eyes and white flesh. The variety is susceptible to late blight disease.
Kufri Pukhraj	70 – 90	35-40	North Indian plains and plateau regions	Plants are tall, semi-erect, medium compact and vigorous and the flower is white in colour. Tubers are white, large, oval, slightly tapered, smooth skin, fleet eyes with waxy textured and yellow coloured flesh. The variety is resistant to early blight and moderately resistant to late blight disease.
Kufri Jawahar	80 – 90	35 -40	Punjab and Haryana, plateau region of Gujarat, Karnataka and Madhya Pradesh	Plants are short, erect, compact and vigorous with light green foliage and white flower. Tubers are creamy white, medium sized, round to oval in shape with smooth skin, shallow deep eyes and pale yellow flesh. It is moderately resistant to late blight disease.
Kufri Kuber	80 – 90 (in plains) and 110 – 135 (in hills)	25-30	North – eastern plains of India	Plants are medium tall, compact, vigorous and spreading habit with dark grey green foliage and white flower. Tubers are medium to large in size, oval shaped, tapering towards crown end and white in colour with medium deep eyes. It is susceptible to early and late blight and wart disease. It is no longer cultivated because of poor keeping quality.

Variety	Duration (days)	Yield (t/ha)	Cultivated area	Characteristics
Kufri Surya	80 – 90	30-35	Early (September) planting in north-western plains; *rabi* and *kharif* plantings in peninsular India; main crop season in north-western and west-central plains	It is a heat tolerant and hopper-burn resistant potato variety with oblong tubers, white smooth skin and pale yellow flesh. It yields excellent defect free tubers with high proportion of large (> 85mm) tubers suitable for processing into high quality French fries and chips. The reducing sugar content of tubers of this variety is less than 100mg/100g fresh weight and the tuber dry matter content is 20–21% at harvest.
Kufri Khyati	80 – 90	25 - 28	Autumn planting in Rajasthan, Haryana, Punjab, Uttar Pradesh, Bihar, West Bengal, Gujarat, Madhya Pradesh, Odisha and Chhattisgarh	Tubers are oval shaped, pale yellow coloured, medium sized with fleet eyes and light yellow flesh. Moderately resistant to late blight.
Kufri Garima	80-90	25-30	North Indian plains and plateau regions	Its canopy is compact, stem predominantly green with red- brown pigment only at base. Leaflet with medium and ovate lanceolate, flowering profuse, corolla white and sprouts purple. It has attractive light yellow, ovoid tubers with shallow eyes and light yellow flesh. Tubers do not show deformities like cracking or hollow heart.
Kufri Gaurav	80-90	25-30	North Indian plains and plateau regions	Its canopy is semi-compact and green stem. Leaflet is ovate-lanceolate, flowers are white and sprouts green. It has white cream, ovoid tubers with medium-deep eyes and white cream flesh. It is nutrient-use efficient variety.

Variety	Duration (days)	Yield (t/ha)	Cultivated area	Characteristics
Medium duration				
Kufri Lalima	90 – 100	35 – 40	Indo-Gangetic plains	Tall, erect, vigorous plants with dark grey green foliage, blue-purple flowers and round uniform shining red, smooth, medium deep eyes and medium – sized tuber. The variety is moderately resistant to early blight.
Kufri Bahar	100 –110	35 – 45	Parts of Haryana, Himachal Pradesh, Jammu and Kashmir and Uttar Pradesh	Tall erect, medium compact and vigorous plants with grey green foliage and the flowers are white in colour. Tubers are white, large, round to oval with medium deep eyes. The variety is susceptible to major diseases but has slow rate of degeneration.
Kufri Sherpa	90 – 110 (in plains) and 110 – 135 (in hills)	20 – 25	Hills of West Bengal and Sikkim	Tall, erect, medium compact and vigorous plants; Tubers are medium in size, round in shape, flattened with white skin and medium deep eyes.
Kufri Sutlej	90 – 100	35 – 40	Bihar, Haryana, Madhya Pradesh, Punjab and Uttar Pradesh	Plants are medium tall, erect, medium compact and vigorous with grey green foliage. Tubers are white, large, oval with smooth skin, fleet eyes and white flesh. It is moderately resistant to late blight disease.
Kufri Anand	100 –110	40 – 45	Plains of Uttar Pradesh and its neighboring states.	Tall, erect, open and vigorous plants with light purple flowers. Tubers are large, oval to oblong, white with fleet eyes. It is resistant to late blight disease.
Kufri Giriraj	110 –130	25 – 30	Hills of Himachal Pradesh; south and north eastern hills	Plants are tall, semi – erect, medium compact and vigorous with light purple colour flowers. Tubers are medium to large, white, oval with fleet eyes and waxy textured flesh. It is field resistant to late blight disease.

Variety	Duration (days)	Yield (t/ha)	Cultivated area	Characteristics
Kufri Kanchan	110 – 130 (during summer) and 90 – 100 (during winter)	25 – 28	North Bengal hills and Sikkim	Plants are medium tall, semi-erect, medium compact, vigorous with purple flowers. Medium to large, oval to oblong with fleet eyed red tubers are produced. The variety is field resistant to late blight and wart disease, and has slow rate of degeneration.
Kufri Red	90 – 120	23 – 26	Utter Pradesh and Punjab	A clonal selection from indigenous variety – Darjeeling Red Round; Tall, erect, vigorous plants with yellow green foliage. Medium sized, round and irregular shaped, red coloured and deep eyed tuber are produced. The variety is susceptible to early blight, late blight and wart disease.
Kufri Muthu	100 – 120 (during summer) and 90 – 100 (during autumn)	23–25	Southern hills of India	Tall, erect, compact and vigorous plants with dark green foliage and white flower. Medium-large, round to oval, white and medium deep eyed tubers are produced. The variety is resistant to late blight but susceptible to early blight.
Kufri Kundan	120 – 135 (in hills and) 90 – 105 (in plains)	12 – 15	North – western hills comprising Himachal Pradesh and hills of Uttar Pradesh	The variety producing tall, erect, medium compact and vigorous plants with grey green foliage and blue -purple flowers with white tips. Tubers are medium, round to oval, flattened and white with medium deep eyes. It can withstand water logging condition better than any other variety, moderately resistant to late blight but susceptible to early blight, wart, charcoal rot and virus diseases.
Kufri Himalini	110-120	27-29	Hill regions and sub-tropical plains of India	Plants are medium-tall, semi-erect, semi-compact and vigorous with light purple flowers. Tubers are medium to large, oval, smooth, white skinned with shallow to slight medium deep eyes and dull while flesh. It is less prone to cracking. It is highly resistant to late blight. It is suitable for growing in the hill regions. The tuber dry matter content is 18.5% at harvest.

Variety	Duration (days)	Yield (t/ha)	Cultivated area	Characteristics
Kufri Shailja	110-120	25-27.5	Hill regions and sub-tropical plains of India	Plants are medium-tall, semi-erect, medium-compact and vigorous with light red-purple flowers. Tubers are medium, round-oval, smooth, white skinned with shallow eyes and pale yellow flesh. The tuber dry matter content is 19.91% at harvest.
Late duration				
Kufri Badshah	100 –120	40 – 45	North Indian plains and Deccan plateau	The plant is tall, erect, medium compact and vigorous with white flower. Tubers are white coloured, oval shaped and large sized with smooth shallow eyes. The variety is resistant to late blight disease and tolerant to frost.
Kufri Swarna	130 – 135 (during summer) and 100 – 110 (during autumn)	25 – 30	South Indian hills	Plant is tall, erect, compact vigorous, with dark green foliage, few and thick stem and the flower is white. Tubers are white in colour, medium size, round to oval shape, smooth skin, fleet eyes with flesh. It is highly resistant to early and late blight diseases and cyst nematodes.
Kufri Khasigaro	100 – 120 (in plains) and 120 – 135 days in hills	25 – 28 (in plains) and 8 – 10 (in hills)	Hilly regions of Assam	The plants are medium tall, slightly spreading compact and vigorous with grey green foliage red-purple coloured flower. Tubers are medium, round to oval, white and medium deep eyed. It possesses a good degree of field resistance to early blight and fairly high degree of resistance to late blight disease but susceptible to wart disease.

Variety	Duration (days)	Yield (t/ha)	Cultivated area	Characteristics
Kufri Jeevan	110 – 120 (in plains) and 130 – 150 (in hills)	17.5 – 20	North – western hills	Plants are tall, erect, medium compact and vigorous with grey green foliage and white flower. Medium to large, oval, white, shallow deep eyed tubers are common. It is field resistant to late blight and wart disease.
Kufri Naveen	130 – 150	20 – 25	North – eastern hills of Assam, West Bengal and Meghalaya	The variety is producing tall, erect, medium compact and vigorous plants with yellow green foliage. Tubers are medium to large in size, round to oval in shape, white in colour with fleet eyes. It is resistant to late blight and wart disease.
Kufri Neelamani	120 –140 (in summer) and 110 – 120 days in autumn	13 – 15 (rainfed condition) and 30 – 35 (irrigated conditions)	Southern hills of India	Plants are tall, erect compact, spreading and vigorous plants with yellow-green foliage and blue-purple flower. Medium, oval to flattened, white and deep eyed tubers are developed. It is moderately resistant to late blight but susceptible to early blight and wart disease.
Kufri Neela	120 – 135 (during summer) and 115 – 125 (during autumn)	12.5 – 15 (rainfed condition) and 25 – 30 (irrigated condition)	Southern hills of India	Plants are tall, erect, medium compact and vigorous with grey-green foliage and blue-purple flower. Tubers are medium, round, white with medium deep eyes. It is moderately resistant to late blight but susceptible to early blight and ward disease.
Kufri Kumar	160 – 165 (in hills) and 110 – 120 (in plains)	25 – 28	North – western hills	Plants are tall, erect, vigorous plants with grey-green foliage and blue-purple flower. Large, oblong, white and fleet eyed tubers are developed. It is moderately resistant to late blight but susceptible to early blight.
For Chips				
Kufri Chipsona 2	90 – 100	32 – 35	Indo – Gangetic plains of Uttar Pradesh, Bihar and West Bengal	Tall, erect, open plants with white flower. Medium to large, round, white and fleet eyed tubers are produced with waxy textured flesh. It is resistant to late blight disease and tolerant to frost. The variety is suitable for preparation of chips.

Variety	Duration (days)	Yield (t/ha)	Cultivated area	Characteristics
Kufri Chipsona 3	110	35-40	Entire North-western and North-central Indian plains	It is a late blight resistant potato variety with round oval tubers, white smooth skin and cream/pale yellow flesh. The variety is meant for processing, especially chip making. The total and process grade tuber yields of Kufri Chipsona-3 are higher than those of Kufri Chipsona-1 and Kufri Chipsona-2. It produces excellent defect-free tubers. The tubers have low reducing sugars (< 50 mg/100 g fresh wt) at harvest and even after six months of storage at 10–12°C. The tuber dry matter content is > 21%.
Kufri Chipsona 4	100 – 110	35-40	West Bengal, Madhya Pradesh and Karnataka.	Its canopy is compact and stem green with red-brown pigment lightly scattered throughout. Leaflet is lanceolate, flowers are white and sprouts red-purple. It has white cream, round tubers with shallow eyes and white flesh. It has high dry matter and low reducing sugars, and thus suitable for making chips. It is field resistant to late blight.
Kufri Frysona	100 – 110	35-40	North Indian plains	Its canopy is open and stem green with purple pigment highly scattered throughout. Leaflet is ovate-lanceolate, flowers are red-violate and sprouts red-purple. It has white cream, long-oblong tubers with shallow eyes and white flesh. It has high dry matter and low reducing sugars, and thus suitable for making chips. It is field resistant to late blight and immune to wart.

References

Agarwal, R. L. 1999. Seed Technology. Oxford & IBH Publishing Co. Pvt. Ltd., pp. 285-289.

Central Potato Research Institute, 1999. Manual for True Potato Seed (TPS) Production and Utilization. Extension Bulletin No. 30 (E).

Central Potato Research Institute. 1987. Extension Bulletin No. 17.

Dissanayake, M.L.B.B. and Devsabai, K. 1988. Result of TPS trials conducted at regional agricultural research centre, Bandasawela. Proceeding of the Regional Workshop on TPS Extension and On-farm Technology Transfer. 22 August, 1988. pp. 7.

International Potato Centre (CIP) Hybrid True Potato Seed Production. IARI, New Delhi, India.

Sharma, R.S. 2002. Text Book of Field Crops Production, Potato. pp. 719-768.

Singh, A.N. 1988. Salient achievements in TPS technology for the last three years in Eastern Indo Gangetic plains (1985-88). Proceeding at Regional Workshop on TPS Extension and On-farm Technology Transfer. 22 August, 1988, pp. 15.

Singh, S. S. 2004. Crop Management. Kalyani Publishers, Ludhiana, India, pp. 305-306.

Index